Studies in Computational Intelligence

Volume 1114

The series "Studies in Computational Intelligence" (SCI) publishes new developments and advances in the various areas of computational intelligence—quickly and with a high quality. The intent is to cover the theory, applications, and design methods of computational intelligence, as embedded in the fields of engineering, computer science, physics and life sciences, as well as the methodologies behind them. The series contains monographs, lecture notes and edited volumes in computational intelligence spanning the areas of neural networks, connectionist systems, genetic algorithms, evolutionary computation, artificial intelligence, cellular automata, self-organizing systems, soft computing, fuzzy systems, and hybrid intelligent systems. Of particular value to both the contributors and the readership are the short publication timeframe and the world-wide distribution, which enable both wide and rapid dissemination of research output.

Indexed by SCOPUS, DBLP, WTI Frankfurt eG, zbMATH, SCImago.

All books published in the series are submitted for consideration in Web of Science.

Said Melliani · Oscar Castillo ·
Abdelmajid El Hajaji
Editors

Applied Mathematics and Modelling in Finance, Marketing and Economics

Springer

Editors
Said Melliani
Department of Mathematics
Sultan Moulay Slimane University
Beni-Mellal, Morocco

Oscar Castillo
Division of Graduate Studies and Research
Tijuana Institute of Technology
Tijuana, Baja California, Mexico

Abdelmajid El Hajaji
National School of Trade and Management
Chouaib Doukkali University
El Jadida, Morocco

ISSN 1860-949X ISSN 1860-9503 (electronic)
Studies in Computational Intelligence
ISBN 978-3-031-42846-3 ISBN 978-3-031-42847-0 (eBook)
https://doi.org/10.1007/978-3-031-42847-0

This Springer imprint is published by the registered company Springer Nature Switzerland AG
The registered company address is: Gewerbestrasse 11, 6330 Cham, Switzerland

Paper in this product is recyclable.

Preface

This book contains the written versions of most of the contributions presented during the first Edition of the International Conference in Applied Mathematics to Finance, Marketing and Economics, which took place at the National School of Commerce and Management in El Jadida, Morocco, from 26 to 27 November 2020.

The meeting provided a setting for discussing recent developments in a wide variety of topics including Mathematical modelling in finance, mathematical Models in Marketing; modelling of financial and economic primitives (interest rates, asset prices, etc.), modelling market behaviour, modelling market imperfections, pricing of financial derivative securities, hedging strategies, numerical methods, financial engineering.

The main goal of the event is to encourage the confident use of applied mathematics and mathematical modelling in finance, economic and the use of Mathematical Models in Marketing. Also to explore and address work at the interface between applied mathematics and applications oriented ideas to the various fields of science. The audience was multidisciplinary allowing the participants to exchange diversified ideas and to show the wide attraction of different topics.

All papers had undergone the careful peer-review before it selected for publications in those volumes.

We believe that this event provided a medium for scientists and experts in the field to effectively communicate and share ideas. We would like to express our sincere thanks to all participants for their contributions and stimulating discussions.

Beni-Mellal, Morocco — Said Melliani
El Jadida, Morocco — Abdelmajid El Hajaji
Tijuana, Mexico — Oscar Castillo

Contents

High-Precision Method for Space-Time-Fractional Klein-Gordon Equation

A. Habjia, A. El Hajaji, J. El Ghordaf, K. Hilal, and A. Charhabil

Abstract This paper presents the space-time fractional Klein-Gordon equations (FKGEs) for the spinless particle in potential field. It defines to describe the Higgs boson and the propagation of a boson in vacuum in Standard Model (SM). Besides, in this paper, the sine method is employed to construct exact solutions of the space-time fractional Klein-Gordon equations. Many new families of exact traveling wave solutions of the space-time fractional Klein-Gordon equations are successfully obtained. It is shown that the proposed method provides a more powerful mathematical tool for solving nonlinear evolution equations in mathematical physics.

1 Introduction

The Higgs-boson particles are the main objects of relativistic quantum mechanics defined by the most basic, Klein-Gordon equation. This makes KG-equation a good ground to show a remarkable unexpected new invariance toward sign-reversal of both the Planck constant $\hbar$ and all "charges" (including electrical charge and gravitation mass) in a particle-antiparticle transformation, replacing thus a standard CPT-invariance for such a process. Because the Higgs boson is a spin-zero particle, it is the first noticed ostensibly elementary particle to be treated by the Klein-Gordon equation (Klein-Fock-Gordon equation or Klein-Gordon-Fock

A. Habjia
LIRST Laboratory, FP, University of Sultan Moulay Slimane, Beni-Mellal, Morocco

A. E. Hajaji (✉)
LERSEM Laboratory, ENCG, University of Chouaïb Doukkali, El Jadida, Morocco
e-mail: a_elhajaji@yahoo.fr

J. E. Ghordaf · K. Hilal
LAMSC Laboratoty, FST, University of Sultan Moulay Slimane, Beni-Mellal, Morocco

A. Charhabil
Paris Sorbonne Nord University, Paris, France
e-mail: charhabil@math.univ-paris13.fr

S. Melliani et al. (eds.), *Applied Mathematics and Modelling in Finance, Marketing and Economics*, Studies in Computational Intelligence 1114,
https://doi.org/10.1007/978-3-031-42847-0_1

equation). The Klein Gordon equation was the first trial to unify special relativity and quantum mechanics. While initially discarded this equation of "many fathers" can be employed to understand spinless particles that, hence, led to the discovery of pions and other subatomic particles. The equation leads to the development of Dirac equation and hence quantum field theory. Yet, with the appropriate interpretation, it does describe the quantum amplitude for detecting a point particle in various places, the relativistic wave function, but the particle propagates both forwards and backwards in time. On July 4, 2012 CERN announced the finding of the Higgs boson. Since the Higgs boson is a spin-zero particle, it is the major elementary particle that is explained by the Klein-Gordon equation. More experiments and detailed examination is needed to find out if the Higgs boson found is that of the Standard Model, or a more exotic form.

The study of Nonlinear Fractional Klein-Gordon equation is very important for both mathematical and the physical applications. Several decades ago, there were notable improvements in the study of exact solutions of nonlinear fractional differential equations; resulting in a variety of methods, Feng et al. developed and study a some Cauchy matrix type solutions for the nonlocal nonlinear equations [20], the exact solution of the fractional differential equation [18], the exact solution of the space-time fractional inhomogeneous nonlinear diffusion equations [19], exact wave solutions for the nonlinear time fractional Sharma-Tasso-Olver equation and the fractional Klein-Gordon equation in mathematical physics [11], Two-Dimensional Differential Transform Method and Modified Differential Transform Method for Solving Nonlinear Fractional Klein-Gordon equation [12], Analytical approach for space-time fractional Klein-Gordon equation [13], A linearized and second-order unconditionally convergent scheme for coupled time fractional Klein-Gordon-Schrödinger equation [14].

Fractional calculus recently, plays a very leading role in many branches of engineering and physical phenomena that arise in mechanics, electrostatics, quantum mechanics, applied particle physics and many other fields of physics. Many effective approaches and powerful methods to obtaining exact solution for the space-time fractional Klein-Gordon and coupled conformable Boussinesq equations using modified extendend Tanh method with Riccati equation (see [4]), explicit solutions and convergence analysis to the time fracional Cahn-Allen and time-fractional Klein-Gordon equations with Riemann-Liouville derivative (see [5]), An efficient numerical scheme to solve fractional diffusion-wave and fractional Klein-Gordon equations (see [6]), analytical solution of time-fractional order Klein-Gordon equations by using a differential transform method with appropriate initial condition (see [7]), in [8] developed a new exact analytical solutions of time-fractional Klein-Gordon equations by means of conformable fractional derivative by using Modified Kudryashov method.

The outline of the present paper is as follows. In Sect. 2, we introduce the sine method and the descripion of devoted to properties of the conformable fractional derivative, and review on space-time fractional Klein-Gordon equations. In Sect. 3, we apply the sine method on the space-time fractional Klein-Gordon equations by means of conformable fractional derivative, respectively. The applications of the above mentioned technique will be illustrated in Sect. 4 with a space-time fractional

Klein-Gordon equations. The sine method will be applied to solve the space-time fractional Klein-Gordon equations, two-dimensional the space-time fractional Klein-Gordon equations and three-dimensional the space-time fractional Klein-Gordon equations also leads to the smoothness parameter, graphical results show the geometric behaviors to the analitical solutions at different values of fraction order and we have interpreted the various cases. Finally a discussion and conclusions is given in Sect. 5.

2 Description of the Fractional Calculus and Sine Method

2.1 Description of the Fractional Calculus

The fractional calculus finds applications in different domains of science, engineering, physics, numerical analysis, biology and economics [15–17]. Through this section we present some mathematical definitions and properties of the conformable fractional derivative calculus (see [15]) which are used in our work.

Definition 1 Definition

Let f: $[a, b] \times (0, \infty) \rightarrow$ R, then the conformable fractional derivative of f is defined as

$$D_t^{\alpha}(f)(x,t) = lim_{\epsilon \to 0} \frac{f(x, t + \epsilon t^{1-\alpha}) - f(x,t)}{\epsilon}, \quad \alpha \in (0, 1], \ \forall t > 0. \tag{1}$$

Every real function which satisfy in Eq. (1) and corresponding limit exist, is called as the $\alpha -$ differentiable function.

Theorem 1 Theorem (see [15]) *Let* $\alpha \in (0, 1]$ *and* $a, b \in$ R, *then*

(i) $D_t^{\alpha}(au + bv) = aD_t^{\alpha}(u) + bD_t^{\alpha}(v)$,

(ii) $D_t^{\alpha}(t^{\lambda}) = \lambda t^{\lambda - \alpha}, \ \lambda \in$ R,

(iii) $D_t^{\alpha}(uv) = uD_t^{\alpha}(v) + vD_t^{\alpha}(u)$,

(iv) $D_t^{\alpha}(\frac{u}{v}) = \frac{uD_t^{\alpha}(v) - vD_t^{\alpha}(u)}{v^2}$,

(v) $D_t^{\alpha}(u) = t^{1-\alpha}u'(t), \ u \in C^1$,

(vi) $D_t^{\alpha}(u \circ v) = t^{1-\alpha}v'(t)u'(v(t))$, *where* $u : (0, \infty) \rightarrow$ R, *is a real differentiable,* $\alpha -$*differentiable function and* v *be a function defined in the range of* u *and also differentiable.*

2.2 Description of Sine Method

We present the sine method by considering the following nonlinear fractional differential equation of the form:

$$P(u, D_t^{\alpha} u, D_{x^i}^{\alpha} u, D_{tt}^{2\alpha} u, D_{x^i x^j}^{2\alpha} u, D_{x^i x^j x^k}^{3\alpha} u, ...) = 0, \quad 0 < \alpha < 1, \tag{2}$$

where D_t^{α} is conformable fractional derivative and $\alpha \in (0, 1)$. Equation (2) has $n+1$ independent variables $(x, t) = (x^1, x^2, ..., x^n, t)$ and one dependent variable $u = u(x, t)$. We will apply simplest equation method [21, 22] to find exact solutions of Eq. (2). Here we outline the main steps of simplest equation method.
Step 1. We introduce the following traveling wave transformation

$$u(x, t) = U(\xi), \ \ \xi = \frac{1}{\Gamma(\alpha + 1)} \sum_{i=1}^{n} l_i (x^i)^{\alpha} - c \frac{t^{\alpha}}{\Gamma(\alpha + 1)} - x_0, \tag{3}$$

where $l_i (i = 1, 2, ...n)$ and c are nonzero constants. Substitution of wave transformation (3) into (2), we obtain an ordinary differential equation of the form

$$Q(U, cU_{\xi}, l_i U_{\xi}, l_i l_j U_{\xi\xi}, l_i l_j l_k U_{\xi\xi\xi}, ...) = 0. \tag{4}$$

where $U_{\xi} = \frac{dU}{d\xi}$, $U_{\xi\xi} = \frac{d^2U}{d\xi^2}$,....
The ordinary differential Eq. (4) is then integrated as long as all terms contain derivatives, where we neglect integration constants.
Step 2. The solutions of many nonlinear equations can be expressed in the form (see [9])

$$U(\xi) = \begin{cases} \lambda \sin^m(\theta\xi), \ |\xi| \le \frac{\pi}{\theta}, \\ 0, \text{ otherwise}, \end{cases} \tag{5}$$

where λ, θ and $m \neq 0$ are are parameters that will be determined, θ and λ are the wave number and the wave speed, respectively. We use

$$\begin{aligned} &U(\xi) = \lambda \sin^m(\theta\xi), \\ &U^n(\xi) = \lambda^n \sin^{nm}(\theta\xi), \\ &U_{\xi}^n = nm\theta\lambda^n \cos(\theta\xi) \sin^{nm-1}(\theta\xi), \\ &U_{\xi\xi}^n = -n^2\theta^2 m^2 \lambda^n \sin^{nm}(\theta\xi) + n\theta^2\lambda^n m(nm - 1) \sin^{nm-2}(\theta\xi). \end{aligned} \tag{6}$$

Step 3. We substitute (6) into the reduced equation obtained above in (4), balance the terms of the sine functions when (6) is used, and solving the resulting system of algebraic equations by using the computerized symbolic calculations. We next collect all terms whit same power in $\sin^k(\theta\xi)$ and set to zero their coefficients to get a system of algebraic equations among the unknowns θ, m and λ. We obtained all possible value of the parameters θ, m and λ (see [9]).

3 Application

The Klein-Gordon equation is a relativistic wave equation linked to the Schrodinger equation. The equation can be formulated by the Schrodinger equation. The equation shows all spinless particles waves functions with positive and negative charge in addition to zero charge. This equation unveils a lot of interesting traveling wave structures that have not yet been found in the past studies. Traveling waves together with their solution expressions in explicit or implicit forms are crucial from the view point of applications. Such types of waves will not change their profiles during propagation and for this reason, they are easily identified. In this section, we establish more general and novel solutions to the nonlinear space-time-fractional Klein-Gordon equation through the sine method to evaluate exact.

3.1 Conformable Space-Time-Fractional Klein-Gordon Equation with Cubic Nonlinearity

In this sub-section, we present relativistic fractional order Klein-Gordon equation with conformable fractional derivative (b_α define as roughness parameter of time). The square root of the parameter b_α is the smoothness parameter for time. Higher mass of particle lowers the smoothness parameters. The fractional K-G equation suggests all the possibilities which can be connected with general theory of relativity.

$$D_{tt}^{2\alpha}u(x,t) - D_{xx}^{2\alpha}u(x,t) - \frac{b_\alpha}{\hbar_\alpha^2}m_\alpha^2c^4u(x,t) - \mu_\alpha u^3(x,t) = 0. \tag{7}$$

with b_α is the roughness fractional parameter of time, $b_\alpha = \frac{\epsilon_\alpha^2}{m_\alpha^2c^4}$ (ϵ_α is the fractional energy), m_α is the fractional mass parameter, $m_\alpha c^2$ is the mass energy, $\hbar_\alpha^2$ is reduced fractional Planck constant and μ_α is the nonlinear fractional parameter.

Making the transformation $a = \frac{b_\alpha}{\hbar_\alpha^2}m_\alpha^2c^4$, and $\mu_\alpha = \mu$, Eq. (7) becomes:

$$D_{tt}^{2\alpha}u(x,t) - D_{xx}^{2\alpha}u(x,t) - au(x,t) - \mu_\alpha u^3(x,t) = 0. \tag{8}$$

On using the wave transformation

$$u(x,t) = U(\xi),\quad \xi = k\frac{x^\alpha}{\Gamma(\alpha+1)} - c\frac{t^\alpha}{\Gamma(\alpha+1)} - x_0,\ (x,t) \in \Omega, \tag{9}$$

we get a reduced ordinary differential equation as follows

$$(k^2 - c^2)U'' + aU + \mu U^3 = 0. \tag{10}$$

3.2 Implementation of Sine-Cosine Method

Substituting (6) into (10) gives

$$\theta^2 m^2 \lambda (c^2 - k^2)\sin^m(\theta\xi) + a\lambda \sin^m(\theta\xi) + (k^2 - c^2)\theta^2\lambda m(m-1)\sin^{m-2}(\theta\xi) + \mu\lambda^3 \sin^{3m}(\theta\xi) = 0. \tag{11}$$

Equating the exponents and the coefficients of each pair of the sine functions, we find the following system of algebraic equations:

$$\begin{aligned}
&(m-1) \neq 0,\\
&m - 2 = 3m,\\
&\lambda m^2(c^2 - k^2)\theta^2 + a\lambda = 0,\\
&\lambda m(m-1)(k^2 - c^2)\theta^2 + \mu\lambda^3 = 0.
\end{aligned}$$

Solving this system yields

$$m = -1, \quad \theta = \pm\sqrt{\frac{-a}{c^2 - k^2}} \quad \text{and} \quad \lambda = \pm\sqrt{\frac{-2a}{\mu}} \tag{12}$$

Consequently, for $\frac{-a}{c^2 - k^2} > 0$, the following periodic solutions:

$$u_1(x,t) = \pm\sqrt{\frac{-2a}{\mu}}\csc\left(\sqrt{\frac{-a}{c^2-k^2}}\xi\right) \text{ where } 0 < \sqrt{\frac{-a}{c^2-k^2}}\xi < \pi, \tag{13}$$

$$\begin{aligned}
u_2(x,t) &= \pm\sqrt{\frac{-2a}{\mu}}\sec\left(\sqrt{\frac{-a}{c^2-k^2}}\xi\right) \text{ where } 0 < |\sqrt{\frac{-a}{c^2-k^2}}\xi| < \pi/2,\\
u_3(x,t) &= \pm\sqrt{\frac{-2a}{\mu}}\operatorname{csch}\left(\sqrt{\frac{-a}{c^2-k^2}}\xi\right) \text{ where } 0 < \sqrt{\frac{-a}{c^2-k^2}}\xi < \pi,\\
u_4(x,t) &= \pm\sqrt{\frac{-2a}{\mu}}\operatorname{sech}\left(\sqrt{\frac{-a}{c^2-k^2}}\xi\right) \text{ where } 0 < |\sqrt{\frac{-a}{c^2-k^2}}\xi| < \pi/2,
\end{aligned}$$

where $\xi = k\frac{x^\alpha}{\Gamma(\alpha+1)} - c\frac{t^\alpha}{\Gamma(\alpha+1)} - x_0$.
For some arbitrary constants a, μ, c, k and α. Working with Mathematica interactively, we proved our solutions are exact.

3.3 Two-Dimensional Space-Time Fractional Klein-Gordon Equations

We consider the (2 + 1)—dimensional (STFKGEs) is in the following form:

$$D_{tt}^{2\alpha}u(x,y,t) - D_{xx}^{2\alpha}u(x,y,t) - D_{yy}^{2\alpha}u(x,y,t) - au(x,y,t) - \mu u^3(x,y,t) = 0. \quad (14)$$

If we use the transformations

$$u(x,y,t) = U(\xi),\ \ \xi = k\frac{x^\alpha}{\Gamma(\alpha+1)} + h\frac{y^\alpha}{\Gamma(\alpha+1)} - c\frac{t^\alpha}{\Gamma(\alpha+1)} - x_0,\ (x,y,t)\in\Omega, \quad (15)$$

we get a reduced ordinary differential equation as follows

$$(k^2+h^2-c^2)U'' + aU + \mu U^3 = 0. \quad (16)$$

Substituting (6) into (16) gives

$$\theta^2 m^2\lambda(c^2-k^2-h^2)\sin^m(\theta\xi) + a\lambda\sin^m(\theta\xi) + (k^2+h^2-c^2)\theta^2\lambda m(m-1)\sin^{m-2}(\theta\xi) + \mu\lambda^3\sin^{3m}(\theta\xi) = 0. \quad (17)$$

Equating the exponents and the coefficients of each pair of the sine functions, we find the following system of algebraic equations:

$$\begin{aligned}
&(m-1)\neq 0,\\
&m-2 = 3m,\\
&\lambda m^2(c^2-k^2-h^2)\theta^2 + a\lambda = 0,\\
&\lambda m(m-1)(k^2+h^2-c^2)\theta^2 + \mu\lambda^3 = 0.
\end{aligned}$$

Solving this system yields

$$m=-1,\quad \theta = \pm\sqrt{\frac{-a}{c^2-k^2-h^2}}\ \text{ and }\ \lambda = \pm\sqrt{\frac{-2a}{\mu}} \quad (18)$$

Consequently, for $\frac{-a}{c^2-k^2-h^2} > 0$, the following periodic solutions:

$$u_1(x,y,t) = \pm\sqrt{\frac{-2a}{\mu}}\csc\left(\sqrt{\frac{-a}{c^2-k^2-h^2}}\xi\right)\ \text{where } 0 < \sqrt{\frac{-a}{c^2-k^2-h^2}}\xi < \pi, \quad (19)$$

$$u_2(x,y,t) = \pm\sqrt{\frac{-2a}{\mu}}\sec\left(\sqrt{\frac{-a}{c^2-k^2-h^2}}\xi\right)\ \text{where } 0 < |\sqrt{\frac{-a}{c^2-k^2-h^2}}\xi| < \pi/2,$$

$$u_3(x,y,t) = \pm\sqrt{\frac{-2a}{\mu}}\operatorname{csch}\left(\sqrt{\frac{-a}{c^2-k^2-h^2}}\xi\right)\ \text{where } 0 < \sqrt{\frac{-a}{c^2-k^2-h^2}}\xi < \pi,$$

$$u_4(x,y,t) = \pm\sqrt{\frac{-2a}{\mu}}\operatorname{sech}\left(\sqrt{\frac{-a}{c^2-k^2-h^2}}\xi\right)\ \text{where } 0 < |\sqrt{\frac{-a}{c^2-k^2-h^2}}\xi| < \pi/2,$$

where $\xi = k\frac{x^\alpha}{\Gamma(\alpha+1)} + h\frac{y^\alpha}{\Gamma(\alpha+1)} - c\frac{t^\alpha}{\Gamma(\alpha+1)} - x_0$.
For some arbitrary constants a, μ, c, k, h and α. Working with Mathematica interactively, we proved our solutions are exact.

3.4 Three-Dimensional Space-Time Fractional Klein-Gordon Equations

We consider the (3 + 1)—dimensional (STFKGEs) is in the following form:

$$D_{tt}^{2\alpha}u(x,y,z,t) - D_{xx}^{2\alpha}u(x,y,z,t) - D_{yy}^{2\alpha}u(x,y,z,t) - D_{zz}^{2\alpha}u(x,y,z,t) - au(x,y,z,t) - \mu u^3(x,y,z,t) = 0. \tag{20}$$

If we use the transformations

$$u(x,y,z,t) = U(\xi),\ \ \xi = k\frac{x^\alpha}{\Gamma(\alpha+1)} + h\frac{y^\alpha}{\Gamma(\alpha+1)} + l\frac{z^\alpha}{\Gamma(\alpha+1)} - c\frac{t^\alpha}{\Gamma(\alpha+1)} - x_0,\ (x,y,z,t)\in\Omega, \tag{21}$$

we get a reduced ordinary differential equation as follows

$$(k^2 + h^2 + l^2 - c^2)U'' + aU + \mu U^3 = 0. \tag{22}$$

Substituting (6) into (21) gives

$$\theta^2 m^2 \lambda(c^2 - k^2 - h^2 - l^2)\sin^m(\theta\xi) + a\lambda\sin^m(\theta\xi) + (k^2 + h^2 + l^2 - c^2)\theta^2\lambda m(m-1)\sin^{m-2}(\theta\xi) + \mu\lambda^3\sin^{3m}(\theta\xi) = 0. \tag{23}$$

Equating the exponents and the coefficients of each pair of the sine functions, we find the following system of algebraic equations:

$$\begin{aligned}
&(m-1) \neq 0,\\
&m - 2 = 3m,\\
&\lambda m^2(c^2 - k^2 - h^2 - l^2)\theta^2 + a\lambda = 0,\\
&\lambda m(m-1)(k^2 + h^2 + l^2 - c^2)\theta^2 + \mu\lambda^3 = 0.
\end{aligned}$$

Solving this system yields

$$m = -1,\quad \theta = \pm\sqrt{\frac{-a}{c^2 - k^2 - h^2 - l^2}}\ \text{ and }\ \lambda = \pm\sqrt{\frac{-2a}{\mu}} \tag{24}$$

Consequently, for $\frac{-a}{c^2 - k^2 - h^2 - l^2} > 0$, the following periodic solutions:

$$u_1(x,y,z,t)=\pm\sqrt{\frac{-2a}{\mu}}csc\left(\sqrt{\frac{-a}{c^2-k^2-h^2-l^2}}\xi\right) \text{ where } 0<\sqrt{\frac{-a}{c^2-k^2-h^2-l^2}}\xi<\pi, \tag{25}$$

$$u_2(x,y,z,t)=\pm\sqrt{\frac{-2a}{\mu}}sec\left(\sqrt{\frac{-a}{c^2-k^2-h^2-l^2}}\xi\right) \text{ where } 0<|\sqrt{\frac{-a}{c^2-k^2-h^2-l^2}}\xi|<\pi/2,$$

$$u_3(x,y,z,t)=\pm\sqrt{\frac{-2a}{\mu}}csch\left(\sqrt{\frac{-a}{c^2-k^2-h^2-l^2}}\xi\right) \text{ where } 0<\sqrt{\frac{-a}{c^2-k^2-h^2-l^2}}\xi<\pi,$$

$$u_4(x,y,z,t)=\pm\sqrt{\frac{-2a}{\mu}}sech\left(\sqrt{\frac{-a}{c^2-k^2-h^2-l^2}}\xi\right) \text{ where } 0<|\sqrt{\frac{-a}{c^2-k^2-h^2-l^2}}\xi|<\pi/2,$$

where $\xi = k\frac{x^\alpha}{\Gamma(\alpha+1)} + h\frac{y^\alpha}{\Gamma(\alpha+1)} + l\frac{z^\alpha}{\Gamma(\alpha+1)} - c\frac{t^\alpha}{\Gamma(\alpha+1)} - x_0$.
For some arbitrary constants a, μ, c, k, h, l and α. Working with Mathematica interactively, we proved our solutions are exact.

4 Numerical Examples and Discussion

We consider the same test-case, in which the model parameters and the model's data are chosen as $a = 1, k = 1, c = -0.5$ and $\mu = -1$. Figures 1 and 2 shows the solutions of the (STFKGEs) equation obtained in our analytic experiments solutions on the interval $[0, 8] \times [0, 10]$. Figure 3 shows the solutions on the interval $[0, 8] \times [0, 8] \times \{0, 5\}$.

Figure 1 suggests that $x \to 0$, the particle is at rest, and consequently, there is no space variation. Then the fractional wave function of Eq. (13) varies with time only. This means that the particle is oscillating in time, but localised in space, which signifies that the particle is localised in space, but becoming older in time. We call this concept a hidden wave. If a particle is localised in space, it does not mean that it is localised in time also. The particle is moving on a timeline. For instance, if we took a particle muon at rest in space i.e. localised with respect to its position, it still decays. This decay says that there is always a dynamic system in time though the particle is at rest (with respect to space). This type of phenomenon can happen due to the porosity or ruggedness of the construction of time. The ruggedness or rugged character is due to the non-zero mass of the particle.

Figure 1 indicates also that the space is homogeneous without any porosity and permeability or roughness. This occurs because the fractional mass of the particle is zero, which signifies that the particle we examine is a photon in free space (vacuum) by means of the special theory of relativity.

Figure 2 shows the physical characteristics of $u(x, y, t = 0)$ corresponding to $\alpha = 0.25, 0.5, 0.75$ and 1.0 respectively. Figure 3 shows the solution behavior of $u(x, y, t = 5)$ for different fraction Brownian motion $\alpha = 0.25, 0.5, 0.75$ and 1.0, we notice the same solution behavior for different fraction Brownian motion, α.

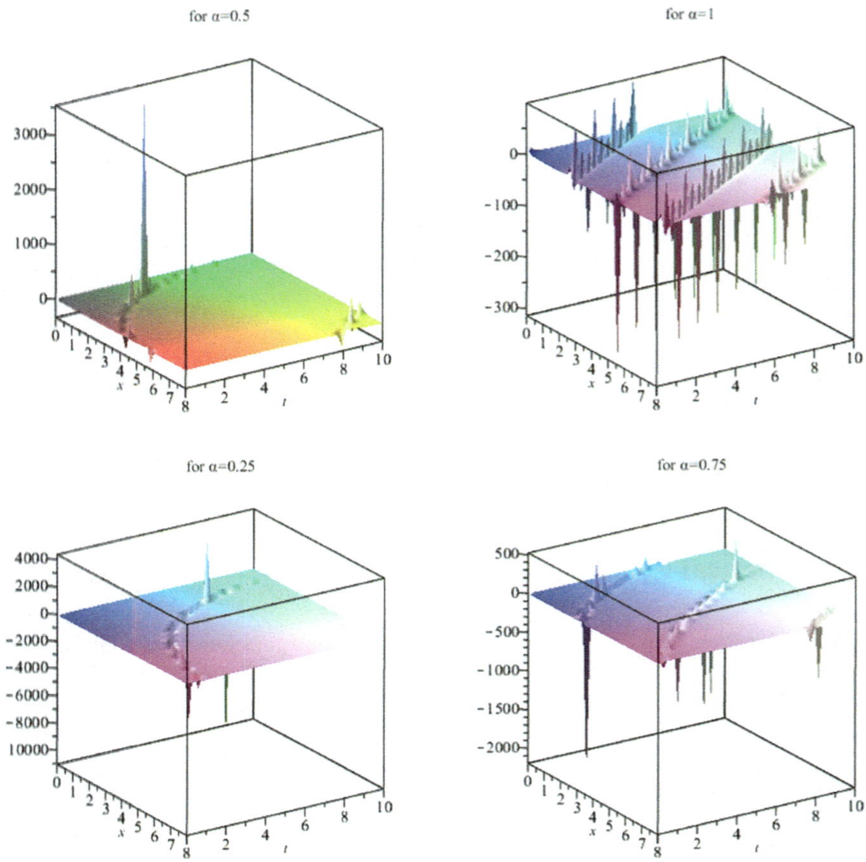

Fig. 1 Oscillating in time of the particle corresponding to $\alpha = 0.25, 0.5, 0.75$ and 1.0 from left to right

This solution for parameter choices $a = 1$, $k = 1$, $c = -0.5$ and $\mu = -1$ is shown in Fig. 2. As can be seen, solution $|u_1|$ given by (19) is the first-order line breather in the (x, y)-plane, which occurs from the constant background possing profiles of parallel lines, and then disintegrated back to the constant background again at great time. The line breather is periodic in both x and y directions. The line breather has the natures: appearing from nowhere and disappear without a trace, which indicates that line blackguard waves may exist in the two dimensional Space-Time-Fractional Klein-Gordon equation (16).

In Figs. 2 and 3, The solution has a line profile with a varying altitude, and is different $(2+1)$-dimensional line solitons. The perfect profile without any disintegrate during their propagation in the (x, y)-plane. Besides, when $t \to +\infty$, this solution $|u_1|$ uniform approaches to the constant background 0; but in the intermediate time,

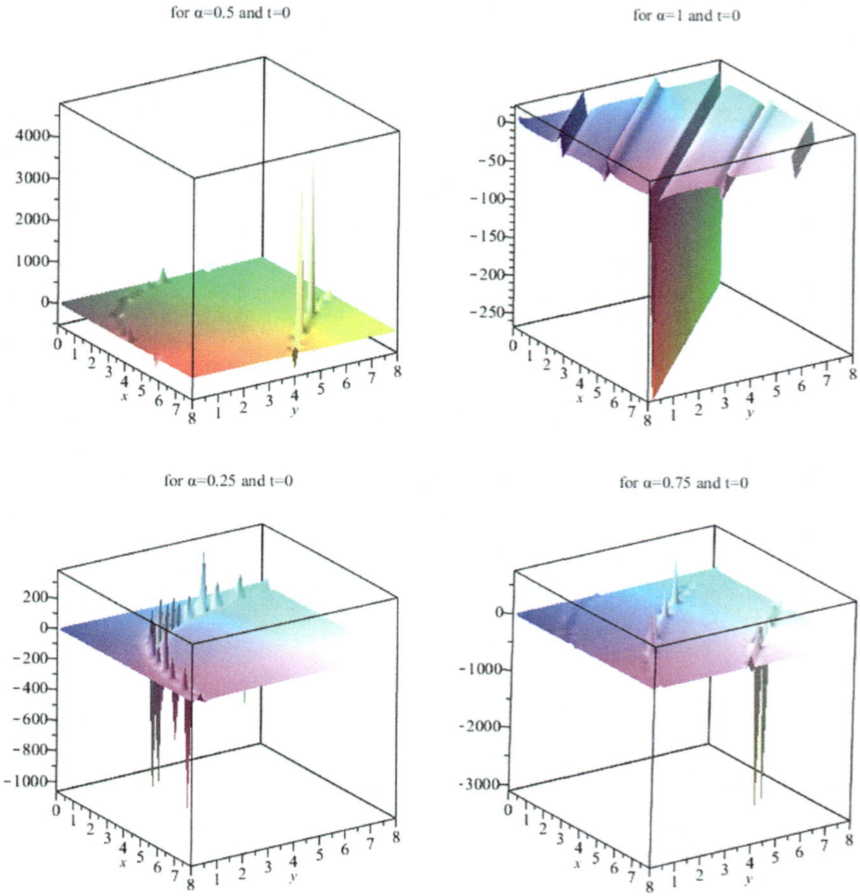

Fig. 2 Physical behavior of $u(x, y, t = 0)$ corresponding to $\alpha = 0.25, 0.5, 0.75$ and 1.0 from left to right

$|u_1|$ attains maximum amplitude 3000 (for $\alpha = 0.5$) at the center of the line wave ($\xi = 0$) at $t = 0$. Hence this line wave describes the phenomenon: line waves appear from anyplace and vanish without a trace. It is famous that the orientation of this line wave is almost arbitrary as the parameter h can be an arbitrary real parameters except 1. In particular, when one takes $h = 0$ in the up line wave, hence the solution u_1 is independent of y. In this case, the two dimensional equation restrains to the one dimensional equation.

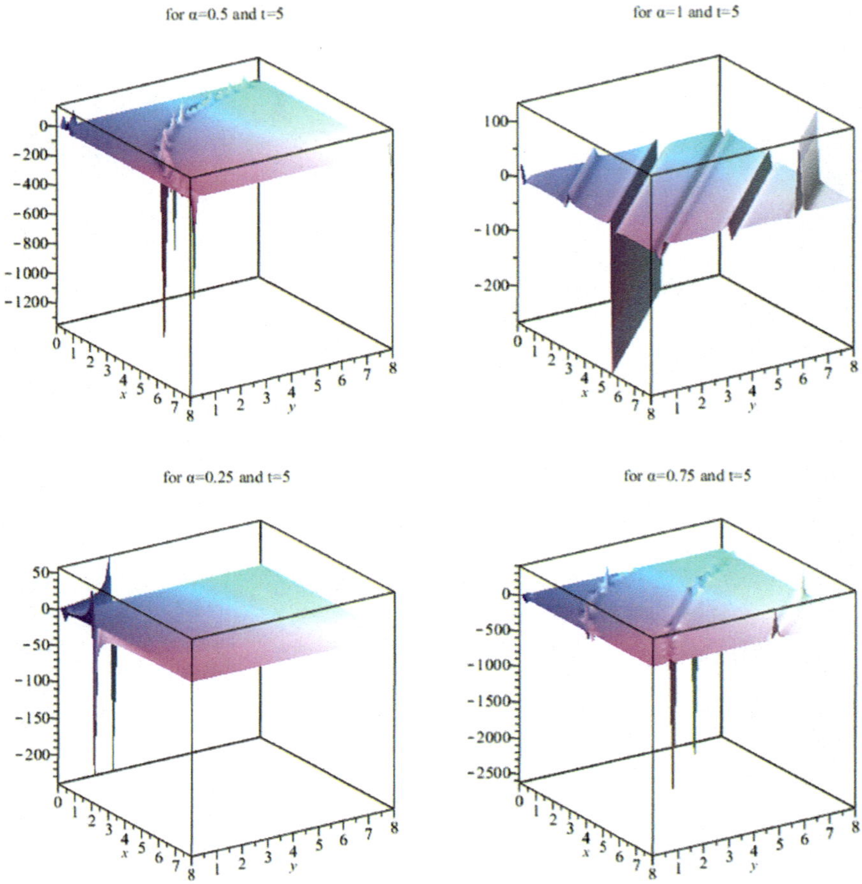

Fig. 3 Physical behavior of $u(x, y, t = 5)$ corresponding to $\alpha = 0.25, 0.5, 0.75$ and 1.0 from left to right

5 Conclusion

The sine method has been successfully utilized to establish exact traveling wave solutions of the space-time fractional Klein-Gordon equations (STFKGEs). The fractional Klein-Gordon equations have various interesting facts like hidden waves, smoothness parameter, negative time-line and a basic connection between our theory and general theory of relativity. The fractional Klein-Gordon equation suggests likewise all the possibilities which can be connected with general theory of relativity. Also, we observed the smoothness parameter which decides the motion of a particle.

These results are going to be very useful invarious areas of applied mathematics such as solid state physics, nonlinear optics, and quantum field theory and others [3]. Finally it is a promising and powerful method for other nonlinear equations in particle physics.

References

1. Khan, T.U., Khan, M.A.: Generalized conformable fractional operators. J. Comput. Appl. Math.**346**, 378–389 (2019)
2. Abdeljawad, T.: On conformable fractional calculus. J. Comput. Appl. Math. **279**, 57–66 (2015)
3. Wazwaz, A.M.: Compactons, solitons and periodic solutions for some forms of nonlinear Klein-Gordon equations. Chaos Solitons Fract. **4**, 1005–1013 (2006)
4. Shallal, M.A., Jabbar, H.N., Ali, K.K.: Analytic solution for the space-time fractional Kein-Gordon and coupled conformable Boussinesq equations. Results Phys. **8**, 372–378 (2018)
5. Inc, M., Yusuf, A., Aliyu, A.I., Baleanu, D.: Time-fractional Cahn-Allen and time-fractional Klein-Gordon equations: Lie symmetry analysis, explicit solutions and convergence analysis. Physica A **493**, 94–106 (2018)
6. Hashemizadeh, E., Ebrahimzadeh, A.: An efficient numerical scheme to solve fractional diffusion-wave and fractional Klein-Gordon equations in fluid mechanics. Physica A **503**, 1189–1203 (2018)
7. Tamsir, M., Srivastava, V.K.: Analytical study of time-fractional order Klein Gordon equation. Alexandria Eng. J. **55**, 561–567 (2016)
8. Hosseini, K., Mayeli, P., Ansari, R.: Modified Kudryashov method for solving the conformable time-fractional Klein-Gordon equations with quadratic and cubic nonlinearities. Optik **130**, 737–742 (2017)
9. Wazwaz, A.M.: A sine-cosine method for handling nonlinear wave equations. Math. Comput. Model **40**, 499–508 (2004)
10. Khalil, R., Horani, M.A., Yousef, A., Sababheh, M.: A new definition of fractional derivative. J. Comput. Appl. Math. **264**, 65–70 (2014)
11. Roy, R., Ali Akbar, M., Wazwaz, A.M.: Exact wave solutions for the nonlinear time fractional Sharma-Tasso-Olver equation and the fractional Klein-Gordon equation in mathematical physics. Opt. Quant. Electron. **50**, 25 (2018)
12. Aruna, K., Ravi Kanth, A.S.V.: Two-Dimensional differential transform method and modified differential transform method for solving nonlinear fractional Klein-Gordon equation. Natl. Acad. Sci. Lett. **37**(2), 163–171 (2014)
13. Unsala, O., Gunerb, O., Bekira, A.: Analytical approach for space-time fractional Klein-Gordon equation. Optik **135**, 337–345 (2017)
14. Lyu, P., Vong, S.: A linearized and second-order unconditionally convergent scheme for coupled time fractional Klein-Gordon-Schrödinger equation. Wiley Periodicals Inc. (2018). https://doi.org/10.1002/num.22282
15. Abdeljawad, T.: On conformable fractional calculus. J. Comput. Appl. Math. **279**, 57–66 (2015)
16. Liu, C.-S.: Counterexamples on Jumarie's two basic fractional calculus formulae. Commun. Nonlinear Sci. Numer. Simul. **22**(1), 924 (2015)
17. Jumarie, G.: Modified Riemann-Liouville derivative and fractional taylor series of nondifferentiable functions further results. Comput. Math. Appl. **51**, 9–10:1367–76 (2006)
18. Bezák, V.: Variations on the linear harmonic oscillator: fourier analysis of a fractional schrodinger equation. Rep. Math. Phys. **84**, No. 3 (2019)
19. Feng, W.: On symmetry groups and conservation laws for space-time fractional inhomogeneous nonlinear diffusion equation. Rep. Math. Phys. **84**(3) (2019)
20. Wei, F., Zhao, S.-L.: Cauchy matrix type solutions for the nonlocal nonlinear Schrödinger equation. Rep. Math. Phys. **84**(1) (2019)

21. Kudryashov, N.A.: Simplest equation method to look for exact solutions of nonlinear differential equations. Chaos Solitons Fract. **24**, 1217–1231 (2005)
22. Kudryashov, N.A.: Exact solitary waves of the Fisher equation. Phys. Lett. A **342**, 99–106 (2005)

Construction of a Bivariate C^2 Septic Quasi-interpolant Using the Blossoming Approach

Abdelhafid Serghini, Abdlemajid El Hajaji, and Ayoub Charhabil

Abstract In this study, we employ a blossoming technique and smoothness criteria to devise a two-step method for creating a C^2 septic spline quasi-interpolant on any given triangulation. This approach ensures an optimal approximation order without the need for coefficient masks associated with smoothness or B-spline basis. To demonstrate the validity of our theoretical findings, we provide numerical experiments.

1 Introduction

The blossoming technique, initially introduced by de Casteljau [8], Ramshaw [27, 28], and other researchers (refer to [13, 20, 21, 39, 40] and related works), proves to be a valuable tool in spline approximations. It offers several advantages, such as the straightforward construction of spline quasi-interpolants in an efficient manner (refer to [7, 26, 30–37, 41] and relevant sources).

In the initial section of this study, we examine certain findings presented in [38], which highlight the relationship between blossoms and splines, particularly in terms of smoothness conditions. The aforementioned publication demonstrates the utility of these results in constructing and analyzing smooth piecewise polynomial functions, including the development of quasi-interpolants.

A. Serghini
ANAA Research Team ESTO, LANO Laboratoty, FSO-ESTO, University Mohammed First, 60050 Oujda, Morocco
e-mail: a.serghini@ump.ac.ma

A. El Hajaji (✉)
LIRO Laboratory, ENCGJ, University Chouaib Doukkali, El Jadida, Morocco
e-mail: a-elhajaji@yahoo.com

A. Charhabil
Paris Sorbonne Nord university, Paris, France
e-mail: charhabil@math.univ-paris13.fr

S. Melliani et al. (eds.), *Applied Mathematics and Modelling in Finance, Marketing and Economics*, Studies in Computational Intelligence 1114,
https://doi.org/10.1007/978-3-031-42847-0_2

The construction of traditional approximations for a given dataset or function typically involves solving linear systems. However, spline quasi-interpolants offer a local behavior that circumvents this issue, making them highly practical. In general, a discrete quasi-interpolant for a given function f, denoted as $\mathcal{Q}f$, is obtained as a linear combination of selected basis functions. The coefficients of this combination correspond to the values of linear functionals that depend on f. Numerous techniques for constructing bivariate discrete quasi-interpolants have been developed and documented in the literature (refer to examples such as [1, 3–6, 9, 15, 29, 38] and relevant references).

Recently, T. Sorokina and Zeilfelder introduced a novel idea in [43] for constructing a C^1 quartic quasi-interpolation $\mathcal{Q}f$ on a uniform three-directional mesh. This method aims to approximate regularly distributed data. In this approach, the B-coefficients of the quasi-interpolant's restriction to each triangle in the partition are determined in a manner that ensures automatic satisfaction of C^1 smoothness. This construction resembles the approach for C^1 quadratic quasi-interpolation presented in [42]. The B-coefficients of $\mathcal{Q}f$ can be readily computed from the given values using local averaging and coefficient masks. It is important to note that the resulting quasi-interpolant is constructed specifically for a particular triangulation and does not span the entire space $\mathbb{P}_4(\mathbb{R}^2)$ of bivariate polynomials of degree less than or equal to 4. It achieves fourth-order accuracy. Recognizing this limitation, a new technique based on blossoming was developed in [38] to define a quasi-interpolation scheme on arbitrary triangulations that is exact on $\mathbb{P}_4(\mathbb{R}^2)$.

In this study, we utilize the same technique presented in [38] to develop an algorithm for constructing a C^2 septic spline quasi-interpolant. Our focus lies in describing the procedure for constructing a C^2 spline quasi-interpolant on arbitrary triangulations, achieving optimal approximation order without the use of coefficient masks for smoothness [42, 43] or B-spline basis [2, 10, 11, 14, 22].

In the proposed algorithm's first stage, a significant portion of the B-coefficients of $\mathcal{Q}f$ are determined by computing the blossoms of local polynomials. This approach automatically satisfies most of the smoothness conditions between the piecewise polynomials of $\mathcal{Q}f$. The remaining B-coefficients around a fixed vertex are obtained by imposing only two smoothness conditions across the edges of the star centered at that vertex. It is important to note that the constructed quasi-interpolant depends on certain parameters, and their selection may not always be straightforward in special cases. In some instances, the C^2 smoothness condition between the first and last adjacent triangles of the closed molecular can be neglected due to the difficulties associated with parameter choices. Consequently, the resulting quasi-interpolant achieves near C^2 smoothness, as the C^2 smoothness condition may not hold for a small number of interior edges.

The paper is structured as follows. In Sect. 2, we provide definitions and properties related to bivariate polynomials and their polar forms. Section 3 revisits relevant findings from [38] regarding the smoothness conditions between polynomials in a spline. Section 4 outlines a two-stage algorithm for constructing a C^2 spline quasi-interpolant of degree 7 on arbitrary triangulations. The algorithm ensures optimal approximation order, and we also discuss the properties of the quasi-interpolation operator. Finally, in Sect. 5, we present several numerical examples to illustrate the theoretical results.

2 Preliminaries

2.1 Polynomials on Triangles

Let's consider a non-degenerate triangle denoted by $\mathscr{T}(V_1, V_2, V_3)$ in the plane, where V_i represents the vertices with Cartesian coordinates (x_i, y_i) for $i = 1, 2, 3$. The barycentric coordinates $\lambda_{\mathscr{T}} := (\lambda_{\mathscr{T},1}, \lambda_{\mathscr{T},2}, \lambda_{\mathscr{T},3})$ of an arbitrary point $(x, y) \in \mathbb{R}^2$ with respect to $\mathscr{T}$ are defined as the unique solution to the system of equations:

$$\begin{bmatrix} x_1 & x_2 & x_3 \\ y_1 & y_2 & y_3 \\ 1 & 1 & 1 \end{bmatrix} \begin{bmatrix} \lambda_{\mathscr{T},1} \\ \lambda_{\mathscr{T},2} \\ \lambda_{\mathscr{T},3} \end{bmatrix} = \begin{bmatrix} x \\ y \\ 1 \end{bmatrix}. \tag{1}$$

We denote by $\mathbb{P}_k(\mathbb{R}^2)$ the linear space of bivariate polynomials of degree less than or equal to k. Set

$$\mathscr{I}_k := \{(i_1, i_2, i_3) \in \mathbb{N}^3, |i| = i_1 + i_2 + i_3 = k\}.$$

Then any polynomial $P \in \mathbb{P}_k(\mathbb{R}^2)$ on $\mathscr{T}$ has a unique representation

$$P(x, y) := b(\lambda_{\mathscr{T}}) = \sum_{i \in \mathscr{I}_k} b_{i,\mathscr{T}} B^{(k)}_{i,\mathscr{T}}(\lambda_{\mathscr{T}}), \tag{2}$$

where

$$B^{(k)}_{i,\mathscr{T}}(\lambda_{\mathscr{T}}) = \frac{m!}{i_1! i_2! i_3!} \lambda^{i_1}_{\mathscr{T},1} \lambda^{i_2}_{\mathscr{T},2} \lambda^{i_3}_{\mathscr{T},3} \tag{3}$$

are the Bernstein-Bézier polynomials of degree k. The coefficients $b_{i,\mathscr{T}}$ are called the Bézier ordinates of the polynomial P with respect to the triangle $\mathscr{T}$.

Consider Ω, a simply connected subset of $\mathbb{R}^2$ with a polygonal boundary denoted by $\partial\Omega$. Let Δ be a triangulation of Ω with vertices V_i having Cartesian coordinates (x_i, y_i) for $i = 1, \ldots, N_v$. Each vertex V_i of Δ corresponds to the set $\mathscr{M}_{V_i}$, which consists of all triangles in Δ that share V_i as a vertex. $\mathscr{M}_{V_i}$ is referred to as the star of Δ centered at V_i.

We define $\mathbb{P}_k^r(\Omega, \Delta)$ as the space of splines of degree k and of class C^r on Ω, i.e.,

$$\mathbb{P}_k^r(\Omega, \Delta) := \{S \in C^r(\Omega) : S_{|\mathcal{T}} \in \mathbb{P}_k, \forall \mathcal{T} \in \Delta\}.$$

2.2 Blossom of a Polynomial

In this subsection, we provide an overview of the fundamental properties of the blossoming principle. The following results can be found in [27, 28] and the related references.

Theorem 1 *Let p be a polynomial in $\mathbb{P}_k(\mathbb{R}^2)$. There exists a unique real function of n variables $X_1, \ldots, X_n$ of $\mathbb{R}^2$, called the blossom of p and denoted by $B[p]$ or $\widehat{p}$, satisfying the following properties:*

1. *Symmetry: $\widehat{p}$ is symmetric for each argument i.e. for any permutation π of $1, 2, \ldots, n$*

$$\widehat{p}(X_1, \ldots, X_n) = \widehat{p}(X_{\pi(1)}, \ldots, X_{\pi(n)}).$$

2. *Affine: $\widehat{p}$ is affine for each argument i.e., if $X_i := \sum_{j=1}^{m} \alpha_j Y_j$ with $\sum_{j=1}^{m} \alpha_j = 1$, then*

$$\widehat{p}(X_1, \ldots, X_{i-1}, X_i, X_{i+1}, \ldots, X_n) = \sum_{j=1}^{m} \alpha_j \widehat{p}(X_1, X_2, \ldots, Y_j, \ldots, X_k).$$

3. *Diagonal: $\hat{p}$ reduces to p when evaluated on its diagonal, i.e.,*

$$\widehat{p}(X, X, \ldots, X) = p(X), \quad \forall X \in \mathbb{R}^2.$$

From [19], we introduce the Leibniz formula which can be used to compute the polar form of a polynomial.

Theorem 2 *Let p_1 and p_2 be two polynomials with degree m and n. If we set $p := p_1 p_2$, we get*

$$\widehat{p}(X_1, \ldots, X_{n+m}) = \frac{\sum_{\pi \in \mathfrak{S}_{m+n}} \widehat{p}_1(X_{\pi(1)}, \ldots, X_{\pi(m)}) \widehat{p}_2(X_{\pi(m+1)}, \ldots, X_{\pi(m+n)})}{(m+n)!}$$

where $\mathfrak{S}_t$ is the symmetric group of all permutations of the set $\{1, \cdots, t\}$.

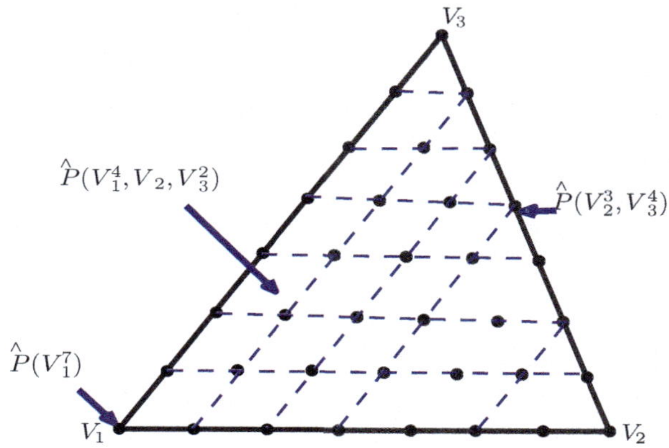

Fig. 1 B-representation of a polynomial P of degree 7, Bézier points are represented by black bullets

It is a well-known fact, as stated on page 14 of [27], that any polynomial p with degree $\leq k$ defined on a triangle $\mathscr{T}(V_1 V_2 V_3)$ can be expressed in the Bernstein basis of $\mathbb{P}_k$ as follows:

$$p(X) = \sum_{i \in \mathscr{I}_k} \widehat{p}(V_1^{i_1}, V_2^{i_2}, V_3^{i_3}) B_{i,\mathscr{T}}^k(\lambda_{\mathscr{T}}(X)), \tag{4}$$

where $\lambda_{\mathscr{T}}(X) := (\lambda_{\mathscr{T},1}(X), \lambda_{\mathscr{T},2}(X), \lambda_{\mathscr{T},3}(X))$ are the barycentric coordinates of the point X with respect to $\mathscr{T}$ and $X^m := \underbrace{(X, \ldots, X)}_{m \text{ times}}$ is in $(\mathbb{R}^2)^m$. The value m is called the multiplicity of X. The following theorem shows that $\widehat{p}(V_1^{i_1}, V_2^{i_2}, V_3^{i_3})$ can be expressed in terms of the values of the restriction of p to the triangle $\mathscr{T}$ at the Bézier points (see Fig. 1)

$$\xi_i := \frac{i_1}{k} V_1 + \frac{i_2}{k} V_2 + \frac{i_3}{k} V_3, \forall i := (i_1, i_2, i_3) \in \mathscr{I}_k. \tag{5}$$

Theorem 3 *Let p be a polynomial of degree $\leq k$ defined on a triangle $\mathscr{T}(V_1 V_2 V_3)$. Then there exists a constant K such that*

$$|\widehat{p}(V_1^{i_1}, V_2^{i_2}, V_3^{i_3})| \leq K \|p\|_{\mathscr{T}}, \forall i \in \mathscr{I}_k$$

Proof Consider the set $\mathscr{D}_k := \xi_i, i \in \mathscr{I}_k$. It has been shown in [12] that the set of nodes $\mathscr{D}_k$ satisfies the geometric characterization (GC) condition. As a result, the Lagrange interpolation at $\mathscr{D}_k$ is uniquely solvable. Thus, any polynomial p with degree $\leq k$ can be expressed as:

$$p(X) = \sum_{i \in \mathscr{I}_k} p(\xi_i) L_i(X),$$

where $L_i, i \in \mathscr{I}_k$ is the Lagrange basis derived from the GC condition. Consequently, for any $j \in \mathscr{I}_k$, we have:

$$\widehat{p}(V_1^{j_1}, V_2^{j_2}, V_3^{j_3}) = \sum_{i \in \mathscr{I}_k} p(\xi_i)\widehat{L}_i(V_1^{j_1}, V_2^{j_2}, V_3^{j_3}).$$

By setting $K = \sum_{i \in \mathscr{I}_k} |\widehat{L}_i(V_1^{j_1}, V_2^{j_2}, V_3^{j_3})|$, we obtain:

$$|\widehat{p}(V_1^{j_1}, V_2^{j_2}, V_3^{j_3})| \leq K|p|_{\mathscr{T}},$$

for all $j \in \mathscr{I}_k$. This result can also be derived using (4) and Theorems 1.11 and 2.4, found on pages 11 and 21 of [25], respectively. ∎

By considering the affinity of the blossom and the like binomial formula given by

$$\begin{aligned} X^m &:= (\lambda_{\mathscr{T},1}(X)V_1 + \lambda_{\mathscr{T},2}(X)V_2 + \lambda_{\mathscr{T},3}(X)V_3)^m \\ &= \sum_{i \in \mathscr{I}_m} B^m_{i,\mathscr{T}}(\lambda_{\mathscr{T}}(X))(V_1^{i_1}, V_2^{i_2}, V_3^{i_3}), \end{aligned} \tag{6}$$

we have the following theorem.

Theorem 4 *Let $\widehat{p}$ be the blossom of a given polynomial $p \in \mathbb{P}_k(\mathbb{R}^2)$. Then for all integer $m \leq k$ and all points $W_1, W_2, \ldots, W_{k-m}$ in $\mathbb{R}^2$ we have:*

(1) $\widehat{p}(W_1, W_2, \ldots, W_{k-m}, X^m)$ is a polynomial in the variable X of degree $\leq m$.
(2) $\widehat{p}$ is affine with respect to argument points of multiplicity m, i.e

$$\begin{aligned} \widehat{p}(W_1, W_2, \ldots, W_{k-m}, X^m) &= \widehat{p}\left(W_1, W_2, \ldots, W_{k-m}, \sum_{i \in \mathscr{I}_m} B^m_{i,\mathscr{T}}(\lambda_{\mathscr{T}}(X))(V_1^{i_1}, V_2^{i_2}, V_3^{i_3})\right) \\ &= \sum_{i \in \mathscr{I}_m} B^m_{i,\mathscr{T}}(\lambda_{\mathscr{T}}(X))\widehat{p}(W_1, \ldots, W_{k-m}, V_1^{i_1}, V_2^{i_2}, V_3^{i_3}) \end{aligned}$$

3 The Blossom of a Spline

In this section, we recall from [38] some interesting definitions and theorems concerning the blossom of a spline or a sub-spline and the smoothness condition between two polynomials. Let $S \in \mathbb{P}_k^0(\Omega, \Delta)$, the blossom of S denoted by $\widehat{S}$ or $B[S]$ is defined as the blossom of each polynomial piece of S i.e.:

$$\hat{S}_{|\mathscr{T} \times \mathscr{T} \times \ldots \times \mathscr{T}} := \hat{S}_{|\mathscr{T}} \qquad \forall\ \mathscr{T} \in \Delta.$$

Then the following theorems derive immediately by replacing the B-coefficients given in Theorem 1 of [24] by the corresponding polar forms.

Theorem 5 *The spline $S \in \mathbb{P}_k^0(\Omega, \Delta)$ is of class C^1 on Ω if and only if $\widehat{S}$ is affine for argument points of multiplicities* 1 *i.e. if the triangle $\widetilde{\mathcal{T}}$ with vertices $\widetilde{V}_1$, V_2, V_3 is adjacent to $\mathcal{T}$ then*

$$\widehat{S}(V_1, V_2^{i_2}, V_3^{i_3}) = \lambda_{\widetilde{\mathcal{T}},1}(V_1)\widehat{S}(\tilde{V}_1, V_2^{i_2}, V_3^{i_3}) + \lambda_{\widetilde{\mathcal{T}},2}(V_1)\widehat{S}(V_2^{i_2+1}, V_3^{i_3}) + \lambda_{\widetilde{\mathcal{T}},3}(V_1)\widehat{S}(V_2^{i_2}, V_3^{i_3+1})$$

for each integers i_2 and i_3 such that $i_2 + i_3 + 1 = k$.

Theorem 6 *The spline $S \in \mathbb{P}_k^0(\Omega, \Delta)$ is of class C^r across the edge V_2V_3 if and only if $\widehat{S}$ is affine for argument points of multiplicities $l \leq r$ i.e.*

$$\widehat{S}(V_1^l, V_2^{i_2}, V_3^{i_3}) = \widehat{S}\left(\sum_{j\in\mathcal{I}_l} B^l_{j,\widetilde{\mathcal{T}}}(\lambda_{\widetilde{\mathcal{T}}}(V_1))(\widetilde{V}_1^{j_1}, V_2^{j_2}, V_3^{j_3}), V_2^{i_2}, V_3^{i_3}\right)$$

$$= \sum_{j\in\mathcal{I}_l} B^l_{j,\widetilde{\mathcal{T}}}(\lambda_{\widetilde{\mathcal{T}}}(V_1))\widehat{S}(\widetilde{V}_1^{j_1}, V_2^{i_2+j_2}, V_3^{i_3+j_3})$$

for each integers i_2 and i_3 such that $i_2 + i_3 + l = k$, $\forall\, l \leq r$ and $\forall \mathcal{T} \in \Delta$. In other words S is of class C^r if and only if $\widehat{S}(V_2^{i_2}, V_3^{i_3}, X^r)$ is a polynomial of degree $\leq r$ on the quadrilateral $Q(V_1V_2V_3\widetilde{V}_1)$ for all edge V_2V_3 in Δ and all integers i_2, i_3 such that $i_2 + i_3 + r = k$ (Figs. 2 and 3).

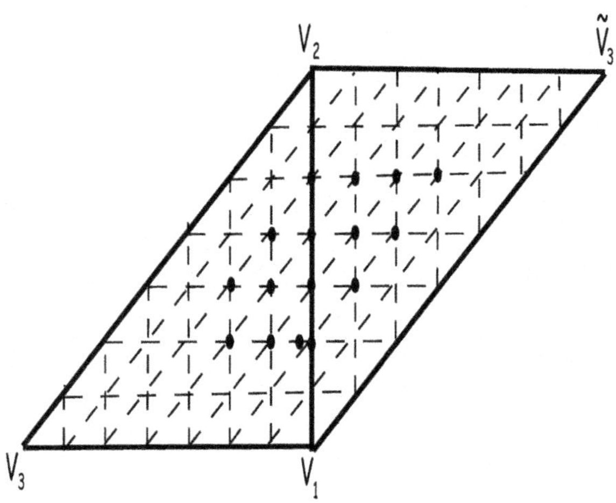

Fig. 2 Localization of the B-coefficients of the sub-spline $\widehat{S}(V_1^2, V_2^2, X^3)$ (bullet points) of a spline of degree 7

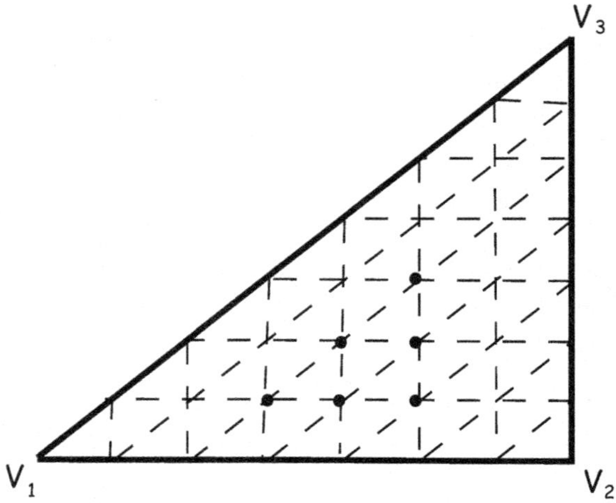

Fig. 3 Localization of the B-coefficients of the sub-spline $\widehat{S}(V_1^2, V_2^2, V_3; X^2)$ of a spline of degree 7 (bullet points)

4 Construction of C^2 Septic Quasi-interpolants

Assuming that certain values of a function f, and possibly its derivatives, are provided at specific data sites within Ω, we utilize the same approach as described in [38] to propose a construction method for a C^2 septic spline quasi-interpolant $\mathcal{Q}f$ of f. The quasi-interpolant $\mathcal{Q}f$ achieves optimal approximation order when the operator $\mathcal{Q}$ accurately represents $\mathbb{P}_7(\mathbb{R}^2)$, i.e., when it is exact on $\mathbb{P}_7(\mathbb{R}^2)$, i.e.,

$$\mathcal{Q}f = f, \quad \forall f \in \mathbb{P}_7(\mathbb{R}^2).$$

4.1 Construction Algorithm

In this subsection, we present the construction method for a C^2 septic spline quasi-interpolant $S := \mathcal{Q}(f)$ defined on an arbitrary triangulation, ensuring optimal approximation order. The majority of the B-coefficients of S are calculated using the blossoming technique applied to local polynomial approximations. The construction is performed through a two-stage algorithm, outlined as follows:

Algorithm 1

Stage 1:
For each triangle $\mathcal{T}$ in the triangulation Δ, we construct a polynomial $p_{\mathcal{T}} := \mathfrak{I}_{\mathcal{T}}(f) \in \mathbb{P}_7(\mathbb{R}^2)$ that serves as a local approximation of f on $\mathcal{T}$. The local approximation operator $\mathfrak{I}\mathcal{T}$ is designed to be exact on $\mathbb{P}_7(\mathbb{R}^2)$ and relies solely on data points within $\mathcal{T}$ or its neighboring triangles.
Stage 2:
Let V be a vertex of the triangulation Δ, and let $\mathcal{M}_V$ represent the star centered at V with its boundary vertices denoted as $V_1, V_2, \ldots, V_{q_v}$, where q_v is the valence of vertex V. We introduce the following notations:

- $\mathcal{T}_j := \mathcal{T}(VV_jV_{j+1}), j = 1, \ldots, q_v$ with $V_{q_v+s} := V_s, \forall s \in \mathbb{N}$.
- $p_j := p_{\mathcal{T}_j}, j = 1, \ldots, q_v$.
- $p_V := \sum_{j=1}^{q_v} \mu_j p_j$, where $\mu_j, j = 1, \ldots, q_v$ are some parameters satisfying $\sum_{j=1}^{q_v} \mu_j = 1$.
- $V_j := \alpha_j V + \beta_j V_{j+1} + \gamma_j V_{j+2}$ with $\alpha_j + \beta_j + \gamma_j = 1, \forall j \in \{1, \ldots, q_v\}$.

Step 1: Set $\widehat{S}(V^4, X^3) := \widehat{p}_V(V^4, X^3), \forall X \in \mathcal{M}_V$. This equality implies that S is C^3 at the vertices.
Step 2: Computation of the B-coefficients $\widehat{S}(V^3, V_j, V_{j+1}^3)$ and $\widehat{S}(V^3, V_{j+1}^3, V_{j+2}), j \in \{1, \ldots, q_v\}$, (see star points in Fig. 4).
Using the smoothness condition across the edge VV_{j+1}, we have

$$\begin{aligned}\widehat{S}(V^3, V_j, V_{j+1}^3) &= \widehat{S}(V^3, \alpha_j V + \beta_j V_{j+1} + \gamma_j V_{j+2}, V_{j+1}^3) \\ &= \alpha_j \widehat{S}(V^4, V_{j+1}^3) + \beta_j \widehat{S}(V^3, V_{j+1}^4) + \gamma_j \widehat{S}(V^3, V_{j+1}^3, V_{j+2}).\end{aligned} \tag{7}$$

If V is an interior vertex, we can choose $\widehat{S}(V^3, V_j, V_{j+1}^3) = \widehat{p}_j(V^3, V_j, V_{j+1}^3)$ and $\widehat{S}(V^3, V_{j+1}^3, V_{j+2})$ is deduced from (7). When V is a boundary vertex, we use the same technique, only for the boundary triangles we have a special treatment, where we can put $\widehat{S}(V^3, V_1^3, V_2) = \widehat{p}_1(V^3, V_1^3, V_2)$ and $\widehat{S}(V^3, V_{q_v-1}, V_{q_v}^3) = \widehat{p}_{q_v-1}(V^3, V_{q_v-1}, V_{q_v}^3)$.
Step 3: Computation of the B-coefficients

$$c_j := \widehat{S}(V^3, V_j^2, V_{j+1}^2), j \in \{1, \ldots, q_v\}, \text{ (see circle points in Fig. 4).}$$

These B-coefficients are computed by C^2 smoothness conditions across the edges $VV_{j+1}, j \in \{1, \ldots, q_v\}$. As $V_j := \alpha_j V + \beta_j V_{j+1} + \gamma_j V_{j+2}$ and by the like-binomial formula

$$V_j^2 = \sum_{i \in \mathcal{I}_2} B^2_{i, \mathcal{T}_{j+1}}(\alpha_j, \beta_j, \gamma_j) V^{i_1} V_{j+1}^{i_2} V_{j+2}^{i_3}$$

we obtain

$$\begin{aligned}c_j &= \sum_{i \in \mathcal{I}_2} B^2_{i, \mathcal{T}_{j+1}}(\alpha_j, \beta_j, \gamma_j) \widehat{S}(V^{i_1+3}, V_{j+1}^{i_2+2}, V_{j+2}^{i_3}) \\ &= B^2_{(0,0,2), \mathcal{T}_{j+1}}(\alpha_j, \beta_j, \gamma_j) \widehat{S}(V^3, V_{j+1}^2, V_{j+2}^2) + d_{j+1} \\ &= \gamma_j^2 c_{j+1} + d_{j+1},\end{aligned}$$

with $d_{j+1} = \sum_{i \in \mathcal{I}_2, i_3 \neq 2} B^2_{i, \mathcal{T}_{j+1}}(\alpha_j, \beta_j, \gamma_j) \widehat{S}(V^{i_1+3}, V_{j+1}^{i_2+2}, V_{j+2}^{i_3})$ and $c_0 = c_{q_v}$ if V is an interior vertex.
Given that the values d_{j+1}, for $j \in 1, \ldots, q_v$, are known (assuming fixed parameters μ_j), the computation of the B-coefficients c_j is performed using the following procedure when V is an interior vertex:

- Since $\prod_{j=1}^{q_v} \gamma_j^2 = 1$, the parameters μ_j are chosen so that $\sum_{j=1}^{q_v} \mu_j = 1$ and $\sum_{i=1}^{q_v} \prod_{s=1}^{i-1} \gamma_s d_{i+1} = 0$, and we take $c_1 = \widehat{p}_1(V^3, V_1^2, V_2^2)$. The others coefficients $c_j, j = 2, \ldots, q_v$ are also deduced from the value of c_1 and Relation (8).

If V is a boundary vertex, we take $c_1 = \widehat{p}_1(V^3, V_1^2, V_2^2)$ and we compute $c_j, j = 2, \ldots, q_{v-1}$ by (8).

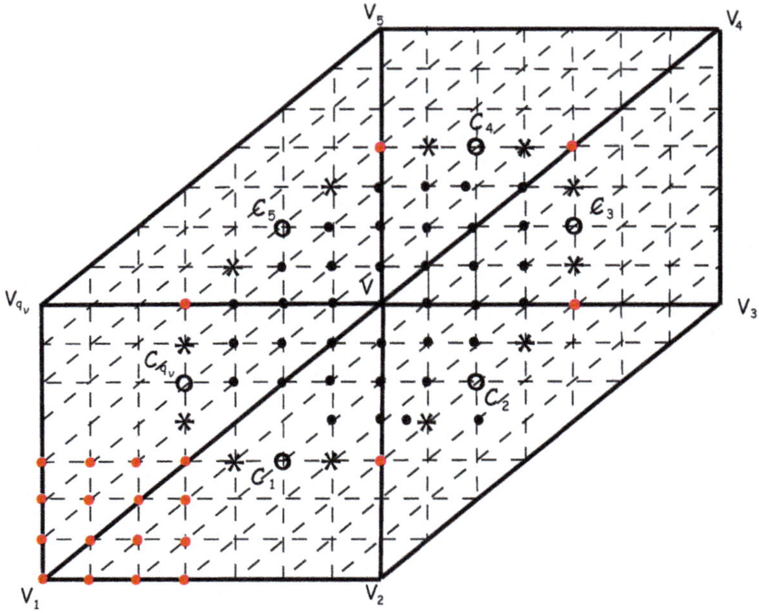

Fig. 4 Localization of the B-coefficients of the quasi-interpolant $\mathcal{Q}f$

Remark 1 Using Theorem 5 and the fact that $\widehat{S}(V^4, X^3)$ is a polynomial of degree 3 (see Step 1 of Stage 2 of Algorithm 4.1), the smoothness conditions between the B-coefficients of $\widehat{S}(V^4, X^3)$, see bullet points in Fig. 4, are implicitly satisfied.

Remark 2 Since $\prod_{j=1}^{q_v} \gamma_j^2 = 1$, it is not easy to choose the parameters μ_j such that $\sum_{j=1}^{q_v} \mu_j = 1$ and $\sum_{i=1}^{q_v} \prod_{s=1}^{i-1} \gamma_s d_{i+1} = 0$, because in this case we lose the local character of the construction and the problem of finding parameters μ_j becomes global, i.e; we must use all μ_j in all interior vertices of Δ. If we neglect this computation, and we take $c_1 = \widehat{p}_1(V^3, V_1^2, V_2^2)$ and we compute c_j, $j = 2, \ldots, q_{v-1}$ by (8), we obtain a quasi-interpolant nearly C^2 (only one C^2 smoothness condition for each interior molecular sub-spline $\widehat{S}(V^3, X^4)$ is not satisfied).

4.2 Properties of the Quasi-Interpolation Operator

Theorem 8 *For any $p \in \mathbb{P}_7(\mathbb{R}^2)$, we have $\mathscr{Q}p = p$.*

Proof The quasi-interpolant $\mathscr{Q}$ being exact on $\mathbb{P}_7(\mathbb{R}^2)$ is guaranteed by two factors. Firstly, the local operators $\mathfrak{I}\mathscr{T}$ are themselves exact on $\mathbb{P}_7(\mathbb{R}^2)$. Secondly, the combination of weights used in the above calculations ensures that their sum is equal to 1. These two factors together ensure the overall exactness of $\mathscr{Q}$ on $\mathbb{P}_7(\mathbb{R}^2)$. ■

Denote by $\Omega_{\mathscr{T}}$ the union of triangles in Δ having a non-empty intersection with $\mathscr{T}$ i.e., $\Omega_{\mathscr{T}} := \mathscr{M}_{V_i} \cup \mathscr{M}_{V_j} \cup \mathscr{M}_{V_k}$, if $\mathscr{T} := V_i V_j V_k \in \Delta$, and put

$$\Delta_{\mathscr{T}} := \{\widetilde{\mathscr{T}} \in \Delta \text{ such that } \mathscr{T} \cap \widetilde{\mathscr{T}} \neq \emptyset\} := \Omega_{\mathscr{T}} \cap \Delta.$$

Suppose that the space $\mathbb{P}_7^2(\Omega, \Delta)$ is equipped with the infinity norm. Then we have the following result.

Theorem 9 *There exists a positive constant K such that*

$$\|\mathscr{Q}f\|_{\mathscr{T}} \leq K \max_{\widetilde{\mathscr{T}} \in \Delta_{\mathscr{T}}} \left\|p_{\widetilde{\mathscr{T}}}\right\|_{\Omega_{\mathscr{T}}} \leq K \max_{\widetilde{\mathscr{T}} \in \Delta_{\mathscr{T}}} \left\|\mathfrak{I}_{\widetilde{\mathscr{T}}}\right\|_{\Omega_{\mathscr{T}}} \|f\|_{\Omega_{\mathscr{T}}}, \tag{9}$$

i.e.,

$$\|\mathscr{Q}\|_{\mathscr{T}} \leq K \max_{\widetilde{\mathscr{T}} \in \Delta_{\mathscr{T}}} \left\|\mathfrak{I}_{\widetilde{\mathscr{T}}}\right\|_{\Omega_{\mathscr{T}}}. \tag{10}$$

Proof The proof is similar that the one of Theorem 17 in [38], we describe it briefly. Let V be a vertex of Δ and $V_1, V_2, \ldots, V_{q_v}$ be the boundary vertices in the star $\mathscr{M}_V$ and put $\mathscr{T}_j := V V_j V_{j+1}$, for a fixed $j \in \{1, \ldots, q_v\}$. Since

$$\|\mathscr{Q}f\|_{\mathscr{T}_j} = \|S\|_{\mathscr{T}_j} \leq \max_{i \in \mathscr{I}_7} \left|\widehat{S}(V^{i_1}, V_j^{i_2}, V_{j+1}^{i_3})\right|,$$

to prove (9), it suffices to demonstrate that there exists a constant K such that for any $i \in \mathscr{I}_7$

$$\left|\widehat{S}(V^{i_1}, V_j^{i_2}, V_{j+1}^{i_3})\right| \leq K \max_{\widetilde{\mathscr{T}} \in \Delta_{\mathscr{T}_j}} \left\|p_{\widetilde{\mathscr{T}}}\right\|_{\Omega_{\mathscr{T}_j}}.$$

On the other hand, we have $\widehat{S}(V^4, X^3) = \widehat{p}_V(V^4, X^3) = \sum_{i=1}^{q_v} \mu_i \widehat{p}_i(V^4, X^3)$, which implies that

$$\left|\widehat{S}(V^4, V_j^{l_2}, V_{j+1}^{l_3})\right| \leq \mu \max_{i=1\ldots,q_v} \left|\widehat{p}_i(V^4, V_j^{l_2}, V_{j+1}^{l_3})\right|,$$

where $l_2, l_3 \in \mathbb{N}$, $l_2 + l_3 = 3$ and $\mu = \sum_{i=1}^{q_v} |\mu_i|$. Using Theorem 3, there exists a constant K_1 such that

$$\left|\widehat{p}_i(V^4, V_j^{l_2}, V_{j+1}^{l_3})\right| \leq K_1 \|p_i\|_{\mathcal{M}_V}.$$

Hence, for $l_2, l_3 \in \mathbb{N}$, such that $l_2 + l_3 = 3$

$$\left|\widehat{S}(V^4, V_j^{l_2}, V_{j+1}^{l_3})\right| \leq K_1 \mu \max_{i=1,\ldots,q_v} \|p_i\|_{\mathcal{M}_V} \leq K_1 \mu \max_{\widetilde{\mathcal{T}} \in \Delta_{\mathcal{T}_j}} \left\|p_{\widetilde{\mathcal{T}}}\right\|_{\Omega_{\mathcal{T}_j}}. \quad (11)$$

Using Step 2 of Algorithm 4.1 and Theorem 3, one can easily see that there exists two constants K_2 and K_3 such that

$$\left|\widehat{S}(V^3, V_j, V_{j+1}^3)\right| \leq K_2 \max_{\widetilde{\mathcal{T}} \in \Delta_{\mathcal{T}_j}} \left\|p_{\widetilde{\mathcal{T}}}\right\|_{\Omega_{\mathcal{T}_j}}.$$

and

$$\left|\widehat{S}(V^3, V_j^3, V_{j+1})\right| \leq K_3 \max_{\widetilde{\mathcal{T}} \in \Delta_{\mathcal{T}_j}} \left\|p_{\widetilde{\mathcal{T}}}\right\|_{\Omega_{\mathcal{T}_j}}.$$

To give an upper bound of $\widehat{S}(V^3, V_j^2, V_{j+1}^2)$, it suffices to use Step 3 of Algorithm 4.1 and Theorem 3. Then, using the same technique given in [38], we can prove that

$$\left|\widehat{S}(V^3, V_j^2, V_{j+1}^2)\right| \leq K_4 \max_{\widetilde{\mathcal{T}} \in \Delta_{\mathcal{T}_j}} \left\|p_{\widetilde{\mathcal{T}}}\right\|_{\Omega_{\mathcal{T}_j}}, \quad (12)$$

where K_4 is a constant.

Hence (9) holds and then (10) is immediately deduced. ∎

Theorem 10 *Let $|\Omega_{\mathcal{T}}| := \max_{u_1,u_2 \in \Omega_{\mathcal{T}}} |u_2 - u_1|$ and assume that the infinity norm of the local approximation operator $\mathfrak{I}_{\mathcal{T}}$ is bounded for any $\mathcal{T} \in \Delta$. Then, there exists a constant K such that for every function $f \in C^{m+1}(\mathbb{R}^2)$, $m = 0, \ldots, 7$*

$$\|\mathcal{Q}f - f\|_{\mathcal{T}} \leq K |\Omega_{\mathcal{T}}|^{m+1} \left\|D^{m+1} f\right\|_{\Omega_{\mathcal{T}}}.$$

Proof Is similar to the prof of Theorem 18 in [38]. ∎

Let $|\Delta|$ denote the maximum diameter among all triangles in the triangulation Δ. If Δ is a uniform triangulation, then we have $|\Delta| = |\mathcal{T}|$ for any triangle $\mathcal{T}$ in Δ. Based on this observation, we can conclude the following result.

Theorem 11 *If Δ is a uniform triangulation and the infinity norm of the local approximation operator $\mathfrak{I}_{\mathcal{T}}$ is bounded for all $\mathcal{T} \in \Delta$. Then, for every function $f \in C^{m+1}(\mathbb{R}^2)$, $m = 0, \ldots, 7$ and $0 \leq |\beta| \leq m$, there exists a constant K_β such that*

$$\left\|D^\beta(\mathcal{Q}f - f)\right\|_{\mathcal{T}} \leq K_\beta |\Delta|^{m+1-|\beta|} \left\|D^{m+1} f\right\|_{\Omega_{\mathcal{T}}}. \quad (13)$$

Proof Is similar to the prof of Theorem 19 in [38]. ∎

5 Numerical Examples

In this section, we present the results of several numerical experiments conducted using our proposed quasi-interpolation approach described in Sect. 4. The experiments were performed using MATLAB with a minimum precision of 15 digits. We considered the well-known uniform three-directional mesh triangulation Δ_l, also known as the type-1 triangulation, of the domain $\Omega = [0, 1] \times [0, 1]$ associated with the vertices (ih, jh), where $i, j = 0, \dots, l$ and $h = \frac{1}{l}$. In our experiments, we chose to simplify the determination of the B-coefficients $c_j = \widehat{S}(V^3, V_j^2, V_{j+1}^2)$ for interior vertices V of Δ. Instead of solving the difficult global problem with $\mu_{i,j}$ as unknowns, we selected the values of $\mu_{i,j}$ without considering the condition $\sum_{i=1}^{q_v} \prod_{s=1}^{i-1} \gamma_s d_{i+1} = 0$ in Step 3 of Algorithm 4.1. As a result, the constructed quasi-interpolant is nearly C^2, meaning that the C^2 smoothness condition is not satisfied at a minority of interior edges. Specifically, out of the eighteen C^2 smoothness conditions for each interior vertex, only one condition is not satisfied. This implies that we achieve 94.44% of C^2 smoothness in our quasi-interpolant.

We use, as bivariate test functions, the bivariate polynomial of degree 7 defined by

$$f_1(x, y) = x^7 - 12xy^6 + 5y^7 + (x - y)(x^2 + y^3 - 1)^2,$$

the Franke's function [18] given by

$$\begin{aligned} f_2(x, y) = {} & \tfrac{3}{4} e^{-((9x-2)^2+(9y-2).^2)/4} + \tfrac{3}{4} e^{-(9x+1)^2/49-(9y+1)/10} \\ & + \tfrac{1}{2} e^{-((9x-7)^2+(9y-3)^2/4)} - \tfrac{1}{5} e^{-(9x-4)^2-(9y-7)^2}. \end{aligned}$$

For different triangulations Δ_l, $l = 8, 16, 32$, we construct our quasi-interpolant defined by Algorithm 4.1.

We estimate the error between f and $\mathcal{Q}f$ by:

$$E(f, \mathcal{Q}f) := \max_{\substack{r=0,\dots,100 \\ s=0,\dots,100}} |f(x_r, y_s) - \mathcal{Q}f(x_r, y_s)|, \text{ where } x_r = \frac{r}{100} \text{ and } y_s = \frac{s}{100}.$$

Table 1 gives the local errors between the septic polynomial f_1 and the quasi-interpolants $\mathcal{Q}f_1$, for the triangulations Δ_l, $l = 8, 16, 32$. It is clear, that the results given in this table confirm that our quasi-interpolant is exact on $\mathbb{P}_7(\mathbb{R}^2)$.

Table 1 Error behavior of Qf_1 for the triangulations Δ_l, $l = 16, 32, 64$

l	8	16	32
$E(f_1, \mathcal{Q}f)$	3.21×10^{-15}	2.66×10^{-15}	2.66×10^{-15}

Table 2 Error between the function f_2 and the septic quasi-interpolants $\mathcal{Q} f_2$, for the triangulation $\Delta_l, l = 16, 32, 64$

l	$\mathcal{Q}_1 f_2$	Order
8	5.68×10^{-4}	7.40
16	3.36×10^{-6}	7.79
32	1.51×10^{-8}	–

Table 2 gives the errors between f_2 and its quasi-interpolant $\mathcal{Q} f_2$. This quasi-interpolant is almost global C^2 over Ω and has lower smoothness C^1 along some edges which are the minority of the interior edges. We also remark that, the approximation ability will not be greatly affected and the approximation order is 8.

6 Conclusion

The work presented in this paper focuses on the construction of a quasi-interpolant spline of degree 7 and class C^2 on arbitrary triangulations. The main advantage of this approach is that it achieves optimal approximation order without the need for coefficient masks for smoothness or B-spline basis. However, it is important to note that the construction of a fully C^2 approximant requires solving a large system, which can be computationally demanding. To address this issue, the paper suggests constructing a nearly C^2 approximant instead. This nearly C^2 approximant provides a good approximation with an acceptable approximation error, while avoiding the computational complexity associated with fully C^2 construction. By relaxing the requirement of full C^2 smoothness, the proposed method offers a practical compromise between accuracy and computational efficiency. The quasi-interpolant retains a high level of smoothness and approximation quality, making it suitable for various applications where a high-order and smooth approximation is desired.

References

1. Abbadi, A., Barrera, D., Ibáñez, M.J., Sbibih, D.: A general method for constructing quasi-interpolants from B-splines. J. Comput. Appl. Math. **234**, 1324–1337 (2010)
2. Alfelf, P., Piper, B., Schumaker, L.L.: An explicit basis for C^1 quartic bivariate splines. SIAM J. Numer. Anal. **24**(4), 891–911 (1987)
3. Barrera, D., Ibáñez, M.J., Sbibih, D., Sablonnière, P.: Near-best quasi-interpolants associated with H-splines on a three-direction mesh. J. Comput. Appl. Math. **183**, 133–152 (2005)
4. Barrera, D., Ibáñez, M.J., Sablonnière, P., Sbibih, D.: On near-best discrete quasi-interpolation on a four-directional mesh. J. Comput. Appl. Math. **233**, 1470–1477 (2010)
5. Barrera, D., Guessab, A., Ibáñez, M.J., Nouisser, O.: Optimal bivariate C^1 cubic quasi-interpolation on a type-2 triangulation. J. Comput. Appl. Math. **234**, 1188–1199 (2010)
6. Barrera, D., Ibáñez, M.J.: Minimizing the quasi-interpolation error for bivariate discrete quasi-interpolants. J. Comput. Appl. Math. **224**, 250–268 (2009)

7. Barry, P.J.: de Boor-Fix, functionals and polar forms. Comput. Aided Geom. Des. **7**, 425–430 (1990)
8. de Casteljau, P.: Shape Mathematics and CAD. Kogan Ltd., London (1985)
9. Chen, G., Chui, C.K., Lai, M.J.: Construction of real-time spline quasi-interpolation schemes. Approx. Theory Appl. **4**, 61–75 (1988)
10. Chui, C.K., Hong, D.: Construction of local C^1 quartic spline elements for optimal-order approximation. Math. Comput. **65**(213), 85–98 (1996)
11. Chui, C.K., Lai, M.J.: Computation of box splines and B-splines on triangulations of non-uniform rectangular partitions. J. Approx. Theory **3**, 37–62 (1987)
12. Chung, K.C., Yao, T.H.: On a lattices admitting unique Lagrange interpolation. SIAM J. Numer. Math. Anal. **14**, 735–743 (1977)
13. Dahmen, W., Micchelli, C.A., Seidel, H.-P.: Blossoming begets B-splines built better by B-patches. Math. Comput. **59**, 97–115 (1992)
14. de Boor, C.: On the evaluation of box splines. Numer. Algoritm. **5**, 5–23 (1993)
15. de Boor, C.: The quasi-interpolant as a tool in elementary polynomial spline theory. In: Berens, H., Cheney, E.W., Lorentz, G.G., Schumaker, L.L. (eds.) Approximation Theory I, pp. 269–276. Academic Press, New York (1973)
16. Farin, G.: Triangular Bernstein-Bézier patches. Comput. Aided Geom. Des. **3**, 19–27 (1986)
17. Farin, G.: Curves and Surfaces for Computer Aided Geometric Design: A Practical Guide, 5th edn. Morgan Kaufmann, San Mateo, CA (2001)
18. Franke, R.: A critical comparison of some methods for interpolation of scattered data, Naval Postgraduate School, Technical report, NPS–53–79–003 (1979)
19. R. Goldman, Blossoming and Divided Difference, Geometric Modelling, Volume 14 of the series Computing, pp. 155–184. Springer Vienna (2001)
20. Gormaz, R.: Floraisons polynomiales: applications à l'étude des B-splines à plusieurs variables. Université Joseph Fourier, Grenoble, Thèse de doctorat (1993)
21. Gormaz, R., Laurent, P.J.: Some results on blossoming and multivariate B-splines. In: Jetter, K., Utreras, F. (eds.) Multivariate Approximation and Wavelets, pp. 147–165. World Scientific, Singapore (1993)
22. Kobbelt, L.: Stable evaluation of box splines. Numer. Algoritm. **14**, 377–382 (1997)
23. Lai, M.J.: A characterisation theorem of multivariate splines in blossoming form. Comput. Aided Geom. Des. **8**(6), 513–521 (1992)
24. Lai, M.J.: Geometric interpretation of smoothness conditions of triangular polynomial patches. Comput. Aided Geom. Des. **14**(2), 191–199 (1997)
25. Lai, M.J., Schumaker, L.L.: Spline Functions on Triangulations. Cambridge University Press (2007)
26. Lamnii, M., Mraoui, H., Tijini, A., Zidna, A.: A normalized basis for C^1 cubic super spline space on Powell-Sabin triangulation. Math. Comput. Simul. **99**, 108–124 (2014)
27. Ramshaw, L.: Blossoming: a connect-the-dots approach to splines, Technical Report 19. Digital Systems Research Center, Palo Alto (1987)
28. Ramshaw, L.: Blossoms are polar forms. Comput. Aided Geom. Des. **6**, 323–358 (1989)
29. Manni, C., Sablonnière, P.: Quadratic spline quasi-interpolants on Powell-Sabin partitions. Adv. Comput. Math. **26**, 283–304 (2007)
30. Sbibih, D., Serghini, A., Tijini, A.: Polar forms and quadratic splines quasi-interpolants over Powell-Sabin triangulation. Appl. Num. Math. **59**, 938–958 (2009)
31. Sbibih, D., Serghini, A., Tijini, A.: Bivariate simplexe spline quasi-interpolants. Numer. Math. Theor. Meth. Appl. **3**(1), 97–118 (2010)
32. Sbibih, D., Serghini, A., Tijini, A.: Normalized trivariate B-splines on Worsey-Piper split and quasi-interpolants. BIT Numer. Math. **52**, 221–249 (2012)
33. Sbibih, D., Serghini, A., Tijini, A.: Superconvergent quadratic spline quasi-interpolants over Powell-Sabin triangulation. Appl. Num. Math. **87**, 74–86 (2015)
34. Sbibih, D., Serghini, A., Tijini, A.: Superconvergent trivariate quadratic spline quasi-interpolants on Worsey-Piper split. J. Comput. Appl. Math. **276**, 117–128 (2015)

35. Sbibih, D., Serghini, A., Tijini, A.: Superconvergent local quasi-interpolants based on special multivariate quadratic spline space over a refined quadrangulation. Appl. Math. Comput. **250**, 145–156 (2015)
36. Sbibih, D., Serghini, A., Tijini, A.: Superconvergent C^1 Cubic Spline Quasi-interpolants on Powell-Sabin Partitions. BIT Numer. Math. **55**(3), 797–821 (2015)
37. Sbibih, D., Serghini, A., Tijini, A.: Trivariate spline quasi-interpolants based on simplex splines and polar forms. Math. Comput. Simul. **118**, 343–359 (2015)
38. Serghini, A., Tijini, A.: New approach to study splines by blossoming method and application to the construction of a bivariate C^1 quartic quasi-interpolant, Comput. Math. Appl. **71**, 529–543 (2016)
39. Seidel, H.P.: Polar forms and triangular B-splines surfaces, in Blossoming: the New polar Form Approach to Spline Curves and Surfaces SIGGRAPH 91, Course Notes #26, ACM SIGGRAPH (1991)
40. Seidel, H.P.: Representing piecewise polynomials as linear combinations of multivariate B-splines. In: Lyche, T., Schumaker, L.L. (eds.) Mathematical Methods in Computer Aided Geometric Design, pp. 559–566. Academic Press, New York (1992)
41. Stefanus, Y., Goldman, R.N.: Blossoming Marsden's identity. Comput. Aided Geom. Des. **9**, 73–84 (1992)
42. Sorokina, T., Zeilfelder, F.: Optimal quasi-interpolation by quadratic C^1 splines on four-directional meshes. In: Chui, C., et al. (eds.) Approximation Theory XI: Gatlinburg 2004, pp. 423–438. Nashboro Press, Brentwood, TN (2005)
43. Sorokina, T., Zeilfelder, F.: An explicit quasi-interpolation scheme based on C^1 quartic splines on type-1 triangulations. Comput. Aided Geom. Des. **25**, 1–13 (2008)

Solving Fuzzy Linear Programming Using the Parametric Form

Abdellatif Semmouri and Mostafa Jourhmane

Abstract The linear programming tool covers a wide range of subject areas including Mathematics, Physics, Financial Management and Digital Economic. Particularly when solving financial planning problems with a goal using linear programming, the presence of fuzziness with the ranking or weighting of goals leads to some technical difficulties. Although significant research works which have been established on fuzzy linear programming, only the membership's aspect or algebraic form are considered. The purpose of this paper is to deal with a kind of linear programming problem involving triangular fuzzy numbers given in the parametric form. In order to demonstrate and to test the proposed methodology, we give an illustrated example. This approach of parametric form will be extended and investigated for solving intuitionistic fuzzy linear programming and neutrosophic linear programming in the future perspective.

Keywords Fuzzy numbers · Parametric form · Ranking function · Fuzzy goal programming

1 Introduction

Linear Programming is a general mathematical framework for modeling and solving some optimization problems. Thus, it becomes a powerful tool for describing and solving linear optimization problems. Mathematically, the problem consists in optimizing a linear function under linear constraints linking certain variables.

Historically, crisp linear programming was first developed and used in 1947 by Danzig [1], Marshall Wood, and their collaborators at the U.S. Department of the Air Force. The first applications were in the military field to fond the optimal military

A. Semmouri (✉) · M. Jourhmane
Faculty of Sciences and Techniques, Sultan Moulay Slimane University, Beni Mellal, Morocco
e-mail: abd.semmouri@gmail.com

M. Jourhmane
e-mail: m.jourhmane@usms.ma

S. Melliani et al. (eds.), *Applied Mathematics and Modelling in Finance, Marketing and Economics*, Studies in Computational Intelligence 1114,
https://doi.org/10.1007/978-3-031-42847-0_3

Fig. 1 George Bernard Danzig (1914–2005)

strategies, but they quickly moved towards industry, allocating resources, scheduling production and workers, planning investment portfolios, formulating marketing and economic planning: for example, it is a question of determining the production maximizing the profit taking into account limited resources or minimizing the costs while guaranteeing a given production, to solve problems of allocation of limited resources in order to achieve set objectives. Nevertheless, there are situations in real life where some parameters of the system are imprecise. However, it is necessary to look for another mathematical discipline which deals with solving some problems related to uncertainty (Fig. 1).

The fuzzy set term first appeared in 1965 [2] when Professor Lotfi A. Zadeh of Berkeley University, USA, published an article entitled "Fuzzy Sets". He has achieved many major theoretical advances in this field and was quickly accompanied by many researchers developing theoretical work. Therefore, many application of fuzzy context have been widely developed and numerous research works have been appeared on development of many aspects of the theory and applications of fuzzy approach. In this context, Zadeh and Bellman extended this theory to solve decision problems because of its efficiency. Then, various works are established in the same literature such that Bellman and Zadeh [3], Ganesan and Veeramani [4], Mesiar et al. [5], Darbari et al. [6] and Zandkarimkhani et al. [7]. In this manuscript, we propose a new approach to solve a fuzzy optimization problem where some parameters are given by α-cuts form.

The structure of this paper is as follows. Section 1 gives a brief historical overview of ordinary linear programming. In Sect. 2, we give some relevant definitions, properties and concepts of fuzzy numbers. Next, Sect. 3 contains the main results in our work. Firstly, we present the fuzzy linear programming model which will be studied in this manuscript. Secondly, we transform this fuzzy problem into a crisp linear programming by the use of a ranking function. In order to demonstrate our result we provide an illustrated example. Finally, the conclusion is suggested in Sect. 4.

2 Preliminaries

In this section, we give some basic concepts, notations and properties related to the fuzzy sets. For more details, interested readers can see Kaufman and Gupta [8], Diamond and Kloeden [9], Klir and Yuan [10], Dubois and Prade [11], Belhallaj and Semmouri [12] and Semmouri et al. [13].

Definition 1 (*Fuzzy set*) Let X be a nonempty base set (universal set) of elements of interest. The fuzzy set $\tilde{A}$ on X (or fuzzy subset of X) is defined by the set of ordered pairs

$$\tilde{A} = \{< x, \mu_{\tilde{A}}(x) >, x \in X\}$$

where $\mu_{\tilde{A}} : X \to [0, 1]$ is a mapping which assigns a real number $\mu_{\tilde{A}}(x)$ taking values in the interval $[0, 1]$ to each element $x \in X$.

For all $x \in X$, $\mu_{\tilde{A}}(x)$ is called the grade (or degree) of membership of x in A.

Definition 2 (α-*cut*) For $\alpha \in]0, 1]$, the α-cut (or α-level as a crisp set) of the fuzzy set $\tilde{u}$ over the universal set X is defined as $\tilde{u}_\alpha := \{x \in X/\mu_{\tilde{u}}(x) \geq \alpha\}$ and $\tilde{u}_0 := cl(\{x \in X/\mu_{\tilde{u}}(x) > 0\})$ where cl(A) denotes the closure of the crisp set A (Fig. 2).

Definition 3 (*Fuzzy number*) A fuzzy set $\tilde{u}$ is called a fuzzy number (FN), if it satisfies the following properties:
(1) $\tilde{u}$ is normal, i.e., there exists $x \in \mathbf{R}$ such that $\mu_{\tilde{u}}(x) = 1$, where $\mathbf{R}$ is the real line.
(2) $\tilde{u}$ is convex, i.e., $\mu_{\tilde{u}}(\lambda x + (1 - \lambda)y) \geq min(\mu_{\tilde{u}}(x), \mu_{\tilde{u}}(y))$, $\forall x, y \in \mathbf{R}$, $\forall \lambda \in [0, 1]$
(3) $\tilde{u}$ is upper semicontinuous.
(4) The α-cut $\tilde{u}_0$ is compact.

The set of all fuzzy numbers is represented by $\mathscr{F}(\mathbf{R})$ and the parametric form of fuzzy numbers is defined in as follows:

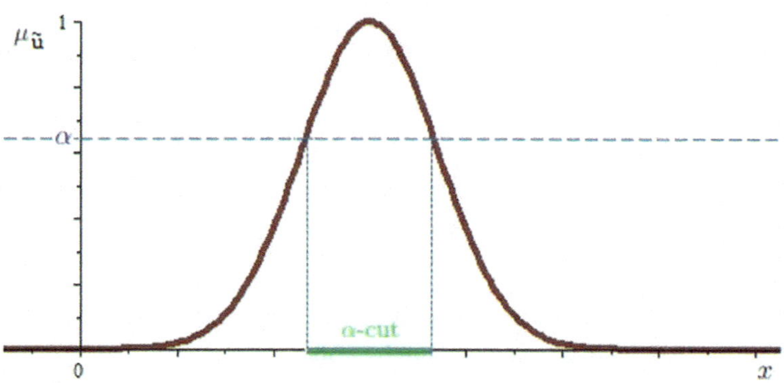

Fig. 2 α-cut of fuzzy set

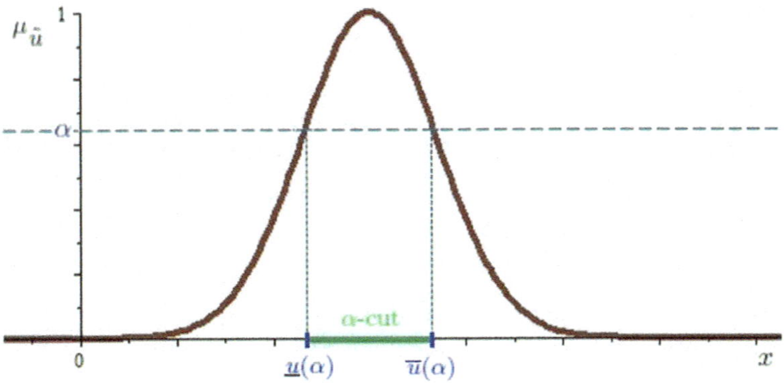

Fig. 3 Parametric form of fuzzy set

Definition 4 (*Parametric form*) A fuzzy number $\tilde{u}$ in parametric form is represented by an ordered pair of functions $(\underline{u}(\alpha), \overline{u}(\alpha))$, $0 \leq \alpha \leq 1$, which satisfy the following requirements (Fig. 3):
(i) $\underline{u}$ is a bounded left-continuous non-decreasing function over $[0, 1]$;
(ii) $\overline{u}$ is a bounded left-continuous non-increasing function over $[0, 1]$;
(iii) $(\underline{u}(\alpha) \leq \overline{u}(\alpha))$, $\forall \alpha \in [0, 1]$.

For arbitrary fuzzy numbers $\tilde{u} = (\underline{u}, \overline{u})$, $\tilde{v} = (\underline{v}, \overline{v})$ and $\lambda \in \mathbf{R}$, we define addition and multiplication by scalar as follows:

$$\underline{u+v} = \underline{u} + \underline{v} \quad and \quad \overline{u+v} = \overline{u} + \overline{v} \tag{1}$$

$$\lambda\tilde{u} = \begin{cases} (\lambda\underline{u}, \lambda\overline{u}) \ if \ \lambda \geq 0 \\ (\lambda\overline{u}, \lambda\underline{u}) \ if \ \lambda \prec 0 \end{cases} \tag{2}$$

Definition 5 (*Triangular fuzzy number*) A triangular fuzzy number $\tilde{a}$ can be defined as a triplet (a_1, a_2, a_3) such that $a_1, a_2, a_3 \in \mathbf{R}$ and $a_1 \leq a_2 \leq a_3$. Its membership function is defined as (Fig. 4):

$$\mu_{\tilde{a}}(x) = \begin{cases} \frac{x-a_1}{a_2-a_1} \ ; \ a_1 \leq x \leq a_2, \\ \frac{a_3-x}{a_3-a_2} \ ; \ a_2 \leq x \leq a_3, \\ 0 \ ; \ otherwise. \end{cases}$$

Obviously, the triangular fuzzy numbers are a particular case of fuzzy numbers.

For arbitrary triangular fuzzy numbers $\tilde{a} = (a_1, a_2, a_3)$, $\tilde{b} = (b_1, b_2, b_3)$ and $\lambda \in \mathbf{R}$, we define addition and multiplication by scalar as follows:

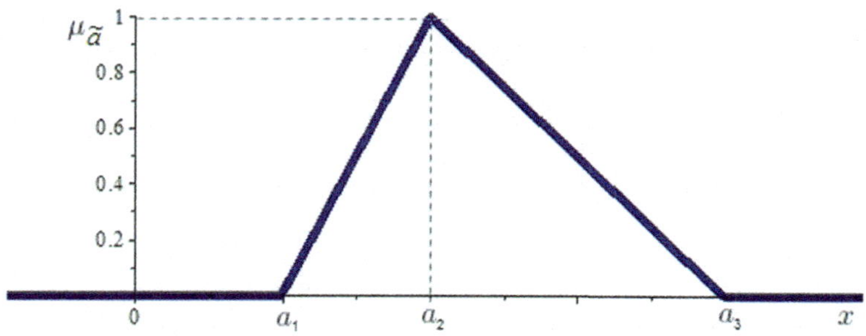

Fig. 4 Triangular fuzzy number

$$\tilde{a} + \tilde{b} = (a_1 + b_1, a_2 + b_2, a_3 + b_3) \tag{3}$$

$$\lambda\tilde{a} = \begin{cases} (\lambda a_1, \lambda a_2, \lambda a_3)\ if\ \lambda \geq 0 \\ (\lambda a_3, \lambda a_2, \lambda a_1)\ if\ \lambda \prec 0 \end{cases} \tag{4}$$

To compare the fuzzy numbers, several points of view were taken. On the one hand, Semmouri et al. [13], Yager [14], Dubois and Prade [15] and Mahdavi-Amiri and Nasseri [16] chose algebraic form or membership function approach for ordering fuzzy quantities. In the other hand, Carlsson and Korhonen [17] and Chutia and Hutia [18] adopt technics based on α-cuts approach.

Definition 6 (*Ordering of FN using the pametric form*) Let $\tilde{u} = (\underline{u}(\alpha), \overline{u}(\alpha))$ and $\tilde{v} = (\underline{v}(\alpha), \overline{v}(\alpha))$, $\alpha \in [0, 1]$ be two FNs in the parametric form. For ranking fuzzy numbers, we define the following pseudo order on $\mathscr{F}(\mathbf{R})$,

$$\tilde{u} \leq_F \tilde{v} \iff \frac{\underline{u}(\alpha) + \overline{u}(\alpha)}{2} \leq \frac{\underline{v}(\alpha) + \overline{v}(\alpha)}{2},\ \forall\alpha \in [0, 1] \tag{5}$$

$$\tilde{u} \geq_F \tilde{v} \iff \tilde{v} \leq_F \tilde{u} \tag{6}$$

In this case, we set $\max(\tilde{u}, \tilde{v}) =_F \tilde{u}$.

$$\tilde{u} \geq_F \tilde{0} \iff \underline{u}(\alpha) + \overline{u}(\alpha) \geq 0,\ \forall\alpha \in [0, 1] \tag{7}$$

3 Results and Discussion

Linear programming succeeded in applied operations research. In the classical approach value of the parameters of linear programming models is well defined and precise. However, in the real environment may be some parameters which are imprecise.

Let us consider fuzzy linear programming with m fuzzy inequality constraints and n fuzzy variables formulated as follows:

$$\max \sum_{j=1}^{n} c_j \tilde{x}_j \tag{8}$$

$$s.t. \quad A\tilde{x} \leq_F \tilde{b} \tag{9}$$

$$\tilde{x} \geq_F \tilde{0} \tag{10}$$

where $\tilde{x} = (\tilde{x}_1, ..., \tilde{x}_n)^T \in (\mathscr{F}(\mathbf{R}))^n$, $A = (a_{ij})$ is a $m \times n$ real matrix, $\tilde{b} = (\tilde{b}_1, ..., \tilde{b}_m)^T \in (\mathscr{F}(\mathbf{R}))^m$ and $c = (c_1, ..., c_n)^T \in \mathbf{R}^n$.

Definition 7 (*Feasible solution*) We say that a fuzzy vector $\tilde{x}$ is feasible solution if and only if (9) and (10) are verified. In this case we say that problem (8)–(10) is feasible.

Definition 8 (*Optimal solution*) The fuzzy optimal solution of fuzzy linear programming will be a fuzzy number $\tilde{x}$ if it satisfies the following characteristics:
(i) $\tilde{x}$ is a feasible solution;
(ii) For each feasible solution $\tilde{x}'$, we have $\sum_{j=1}^{n} c_j \tilde{x}'_j \leq_F \sum_{j=1}^{n} c_j \tilde{x}_j$.

For $\alpha \in [0, 1]$, consider the following crisp linear programming (LP_α)

$$\max \sum_{j=1}^{n} c_j (\underline{\tilde{x}_j}(\alpha) + \overline{\tilde{x}_j}(\alpha)) \tag{11}$$

$$s.t. \quad \sum_{j=1}^{n} a_{ij} [\overline{\tilde{x}}_j(\alpha) + \underline{\tilde{x}}_j(\alpha)] \leq \underline{\tilde{b}}_i(\alpha) + \overline{\tilde{b}}_i(\alpha) \qquad i = 1, ..., m \tag{12}$$

$$\underline{\tilde{x}_j}(\alpha) \leq \overline{\tilde{x}_j}(\alpha) \geq 0 \qquad j = 1, ..., n \tag{13}$$

$$\underline{\tilde{x}_j}(\alpha) + \overline{\tilde{x}_j}(\alpha) \geq 0 \qquad j = 1, ..., n \tag{14}$$

Lemma 1 *The problem (8)–(10) is feasible if and only if the problem (LP_α) is feasible for all $\alpha \in [0, 1]$.*

Proof If problem (8)–(10) is feasible, then (9) and (10) hold. Using (1)–(2) and (5)–(7), we show that problem (LP_α) is feasible for all $\alpha \in [0, 1]$. The converse is obvious.

Combining (1)–(2), (5)–(7) and the previous lemma, we get the following theorem which links between fuzzy optimization and crisp optimization.

Theorem 1 *The problem (8)–(10) is solvable if and only if the problem* (LP_α) *is solvable for all* $\alpha \in [0, 1]$

In practical case, we discretize the interval [0, 1] by introducing a uniformly partitioned mesh. The points in the mesh are $\alpha_k = k\Delta\alpha, k = 0, 1, ..., K$, where $\Delta\alpha = \frac{1}{K}$ is the constant length of the interval steps.

Let $\tilde{x}^\alpha = (\tilde{x}_1^\alpha, ..., \tilde{x}_n^\alpha)^T \in (\mathscr{F}(\mathbf{R}))^n$ be an extreme point solution of the linear programming (LP_α).

In numerical calculation, the character of a fuzzy number requires that:

$$\underline{\tilde{x}}_j^{\alpha_k} \leq \underline{\tilde{x}}_j^{\alpha_{k+1}} \text{ and } \overline{\tilde{x}}_j^{\alpha_{k+1}} \leq \overline{\tilde{x}}_j^{\alpha_k} \qquad \text{for } j = 1, ..., n, k = 0, 1, ..., K$$

By re-bonding of the elementary solutions of the linear programming problem $(LP_{\alpha_k}), k = 0, 1, ..., K$ we get une approximate fuzzy solution of the original linear programming problem (8)–(10).

Example 1 Consider the following fuzzy linear programming

$$\max \tilde{x}_1 + \tilde{x}_2 + \tilde{x}_3$$
$$\text{s.t} \begin{cases} 2\tilde{x}_1 + 3\tilde{x}_2 + \tilde{x}_3 \leq \tilde{b}_1 \\ 4\tilde{x}_1 - 5\tilde{x}_2 + 2\tilde{x}_3 \leq \tilde{b}_2 \\ \tilde{x}_1, \tilde{x}_2, \tilde{x}_3 \geq_F \tilde{0} \end{cases}$$

where

$$\tilde{b}_1 = (\underline{b}_1(\alpha) = 2 + \alpha, \overline{b}_1(\alpha) = 4 - \alpha), \quad \alpha \in [0, 1]$$

and

$$\tilde{b}_2 = (\underline{b}_2(\alpha) = 3 + \alpha, \overline{b}_2(\alpha) = 6 - 2\alpha), \quad \alpha \in [0, 1]$$

We transform the previous fuzzy linear program into the crisp linear program (LP_α) using the pametric form:

$$\max \underline{x}_1 + \overline{x}_1 + \underline{x}_2 + \overline{x}_2 + \underline{x}_3 + \overline{x}_3$$
$$\text{s.t} \begin{cases} 2(\underline{x}_1 + \overline{x}_1) + 3(\underline{x}_2 + \overline{x}_2) + (\underline{x}_3 + \overline{x}_3) = 6 \\ 4(\underline{x}_1 + \overline{x}_1) - 5(\underline{x}_2 + \overline{x}_2) + 2(\underline{x}_3 + \overline{x}_3) = 9 - \alpha \\ \underline{x}_1 \leq \overline{x}_1 \\ \underline{x}_2 \leq \overline{x}_2 \\ \underline{x}_3 \leq \overline{x}_3 \\ \underline{x}_1 + \overline{x}_1 \geq 0 \\ \underline{x}_2 + \overline{x}_2 \geq 0 \\ \underline{x}_3 + \overline{x}_3 \geq 0 \end{cases}$$

By applying some soft optimization program such that *AMPL*, *CPLEX*, *Excel Solvor*, we get the following results (Figs. 5, 6 and Table 1):

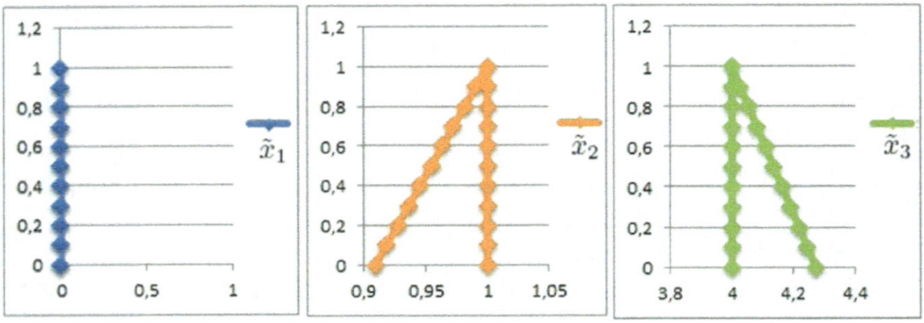

Fig. 5 Membership functions of the fuzzy solutions

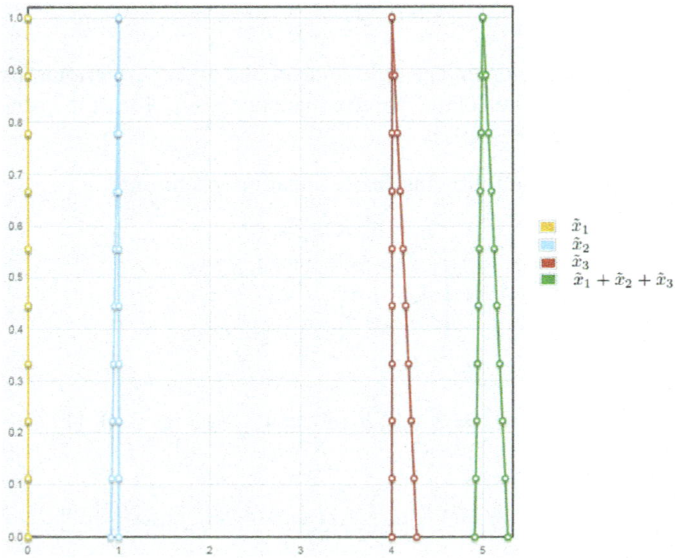

Fig. 6 Membership functions of the fuzzy solutions and the fuzzy maximum

4 Conclusion and Future Perspective

In recent years, fuzzy linear programming has achieved considerable success. In this context, the efforts of researchers are constantly increasing to establish results compatible with the needs of real life. On our side, we have proposed a new method to solve linear programming problems that require the tool of fuzzy theory based on the α-cuts approach. By transforming the linear fuzzy program into ordinary linear programs by using a ranking function for ordering fuzzy numbers. It is a defuzzification technique.

Table 1 Excel solver results

α	$\underline{x}_1$	$\overline{x}_1$	$\underline{x}_2$	$\overline{x}_2$	$\underline{x}_3$	$\overline{x}_3$
0.0	0	0	0.9091	1	4	4.2727
0.1	0	0	0.9182	1	4	4.2455
0.2	0	0	0.9273	1	4	4.2182
0.3	0	0	0.9364	1	4	4.1909
0.4	0	0	0.9455	1	4	4.164
0.5	0	0	0.9545	1	4	4.1364
0.6	0	0	0.9636	1	4	4.1091
0.7	0	0	0.9727	1	4	4.0818
0.8	0	0	0.9818	1	4	4.0545
0.9	0	0	0.9901	1	4	4.0273
1.0	0	0	1	1	4	4

Thus, the numerical resolution of the new programs requires a discretization of the interval [0,1]. The solution of the original problem is a reconstruction of the membership functions from the solutions of the solved sub-problems. It is therefore a fuzzification of the obtained crisp numérical results.

In the future works, we will extend the previous techniques for solving fuzzy linear programming where some parameters are represented by intuitionistic fuzzy numbers.

Acknowledgements The authors would like to thank the following people. Firstly, Professor Dr. S. Melliani of Sultan Moulay Slimane University, Beni Mellal, Morocco for his help and encouraging. Secondly, Mr. Lekbir Tansaoui, ELT teacher, co-author and textbook designer in Mokhtar Essoussi High School, Oued Zem, Morocco for proofreading this paper. We also wish to express our sincere thanks to all members of the organizing committee of the Conference ICAMFME'2020 and referees for careful reading of the manuscript, valuable suggestions and of a number of helpful remarks.

References

1. Dantzig, G.B.: I Complete Form of the Neyman-Pearson Lemma; II On the Non-Existence of Tests of "Student's" Hypothesis Having Power Functions Independent of Sigma. Doctoral dissertation, Department of Mathematics, University of California at Berkeley (1946). [QA276.D3 at Math-Stat Library, University of California, Berkeley.]
2. Zadeh, L.A.: Fuzzy sets. Inf. Control **8**, 338–353 (1965)
3. Bellman, R.E., Zadeh, L.A.: Decision making in a fuzzy environment. Manag. Sci. **17**, 141–164 (1970)
4. Ganesan, K., Veeramani, P.: Fuzzy linear programs with trapezoidal fuzzy numbers. Ann. Oper. Res. **143**, 305–315 (2006)
5. Mesiar, R., Kouchakinejad, F., Šipošová, A.: On fuzzy solution of a linear optimization problem with max-aggregation function relation inequality constraints. Ann. Oper. Res. **269**(1–2), 521–533 (2018)

6. Darbari, J.D., Kannan, D., Agarwal, V., Jha, P.C.: Fuzzy criteria programming approach for optimising the TBL performance of closed loop supply chain network design problem. Ann. Oper. Res. **273**(1–2), 693–738 (2019)
7. Zandkarimkhani, S., Mina, H., Biuki, M., Govindan, K.: A chance constrained fuzzy goal programming approach for perishable pharmaceutical supply chain network design. Ann. Oper. Res. **295**(1), 425–452 (2020)
8. Kaufman, A., Gupta, M.M.: Introduction to Fuzzy Arithmetic, Theory and Applications. Van Nostrand, New York (1985)
9. Diamond, P., Kloeden, P.: Metric Spaces of Fuzzy Sets, Theory and Applications. World Scientific, Singapore (1994)
10. Klir, G., Yuan, B.: Fuzzy Sets and Fuzzy Logic, Theory and Applications. Prentice Hall, PTR, NJ (1995)
11. Dubois, D., Prade, H.: Fundamentals of Fuzzy Sets. Kluwer Academic Publishers, Boston (2000)
12. Belhallaj, Z., Semmouri, A.: Solving fuzzy systems of linear equations by the First-Order Richardson Method. In: 2019 International Conference of Computer Science and Renewable Energies (ICCSRE), pp. 1–5. IEEE (2019)
13. Semmouri, A., Jourhmane, M., Belhallaj, Z.: Discounted Markov decision processes with fuzzy costs. Ann. Oper. Res. **295**(2), 769–786 (2020)
14. Yager, R.R.: A procedure for ordering fuzzy subsets of the unit interval. Inf. Sci. **24**, 143–161 (1981)
15. Dubois, D., Prade, H.: Ranking of fuzzy numbers in the setting of possibility theory. Inf. Sci. **30**(3), 183–224 (1983)
16. Mahdavi-Amiri, N., Nasseri, S.H.: Duality in fuzzy number linear programming by use of a certain linear ranking function. Appl. Math. Comput. **180**, 206–216 (2006)
17. Carlsson, C., Korhonen, P.: A parametric form approach to fuzzy linear programming. Fuzzy Sets Syst. **20**, 17–30 (1986)
18. Chutia, R., Hutia, B.: A new method of ranking parametric form of fuzzy number using value and ambiguity. Appl. Soft Comput. **52**, 1154–1168 (2017)

Dynamic and Static Simulated Annealing for Solving the Multi-objective k-Minimum Spanning Tree Problem

El Houcine Addou, Abdelhafid Serghini, and El Bekkaye Mermri

Abstract This paper deals with the optimisation of the Multi-Objectif k-Minimum Spanning Tree (MO k-MST) problem. A wide varieties of decision making problems in the real world can be formulated as a MO k-MST, which is known to be NP-complete. In order to solve a such problem, we propose two approximate approaches based on simulated annealing method: the first one will integrates the static weighted sum method while the second one uses the dynamic weighted sum method. Computational experiments were carried out in order to compare the performance of each method.

1 Introduction

Simulated Annealing is a probabilistic technique, which is widely used to solve combinatorial optimization problems. In this work we interest in solving the MO k-MST, which is a generalization of the well-known MO MST problem, where several weights are assigned to each edge, this means that several objectives have to be optimized at the same time, and the solution tree must have only k edges, In practice, the objectives are often contradictory, an improvement of an objective can be obtained only at the expense of the other [15], the real solutions are a set of Pareto optimal solutions [4], However, the calculation of these solutions is a hard task because the problem is NP-hard. If we neglect the cardinality k of the problem, i.e. the MO MST has been the subject of several researches thanks to its wide appli-

E. H. Addou (✉) · A. Serghini
ANAA Research Team, ESTO, Laboratory LANO, FSO, University Mohammed Premier, Oujda, Morocco
e-mail: e.addou@ump.ac.ma

A. Serghini
e-mail: a.serghini@ump.ma

E. B. Mermri
Department of Mathematics, FSO, University Mohammed Premier, Oujda, Morocco
e-mail: e.mermri@ump.ac.ma

S. Melliani et al. (eds.), *Applied Mathematics and Modelling in Finance, Marketing and Economics*, Studies in Computational Intelligence 1114,
https://doi.org/10.1007/978-3-031-42847-0_4

cability in many real-world problems (transportation, network design, etc). In literature, many efforts have been contributed to solve the problem with the development and application of a several of optimization algorithms, such as genetic algorithms [5, 9, 11, 18], particle swarm optimization [7, 8], the Greedy Randomized Adaptive Search Procedure [1], and so on. However, if we consider the cardinality k, i.e. the MO k-MST, we notice that no previous work was reported in the literature, which encouraged us to present the problem formulation and provide some approaches to deal with it. In the literature, two weighted sum methods were developed in order to transform the multi-objective optimization problems to a mono-objective one:

- The static weighted sum method [14], in this method, which is also called classical weighted sum method, the weight values assigned to each objective are a static values during the whole algorithm.
- The dynamic weighted sum method, in this approach, the weight values are updated automatically in order to ensure an equitable treatment of each objective function, this method seems to be very effective for the multi-objective optimization problems [6].

The remaining of the paper is organized as follows. We introduce the problem in Sect. 2. In Sect. 3, we propose two solutions method based on SA algorithm, the dynamic sum weighted is used in the first approach, whereas the classical weighted sum method is used in the second approach. Section 4 provides the experimental numerical results. Finally, in Sect. 5, we give some conclusions.

2 Problem Formulation

Given a weighted and undirected graph G with a set of vertices V and a set of edges E, A k-MST is a subtree of G which fulfills the following conditions:

- The subtree contains only k edges where $k <= |V| - 1$.
- The subtree is connected and don't contains any cycles.
- The sum of its edges cost is minimal.

The k-MST problem can be formulated as follows:

$$Minimize f(x) = \sum_{e \in E(T_k)} w(e) \tag{1}$$

subject to $T_k \in X_k$.

The $E(T_k)$ is the edges set of T_k and $w(e)$ denotes the weight (or cost) assigned to the edge e.

Multi-objective optimization addresses the optimization problems with discrete variables and multiple objectives to be optimized at the same time [13], the problem is NP-hard. Therefore, we need to develop efficient approximate algorithms based on heuristics in order to find optimal solutions in reasonable time, the multi-objective optimization can be expressed as follows:

$$Minimize(f_1(x), f_2(x), ..., f_m(x))^I \tag{2}$$

subject to $x \in D$.
where $x = (x_1, x_2, ..., x_n)^L$ is a n-dimensional feasible solution, $(f_1(x), f_2(x), ..., f * m(x))^I$ is a m-dimensional objective space, $D \subseteq R^n$ is a n-dimensional decision space, and m $(m \geq 2)$ is the number of objectives. In this paper we use the weighted sum method which transforms the multi-objective optimization problem (2) into a mono-objective problem as follows [6]:

$$MinimiseZ(x) = \sum_{i=1}^{m} w_i f_i(x) \tag{3}$$

where $1 \leq i \leq m, 0 \leq w_i \leq 1$ and $\sum_{i=1}^{m} w_i = 1$.

The choice of weights w_i is a hard task for the system analyzer and also for decision maker, because the choice must guarantee an equitable treatment of all objectives, hence, the problem is to find the right weights value that characterize the decision makers preferences.

In this work, we suggest two approaches to characterize the weight w_i :

- Classical sum weighted method: In this approach, the value w_i assigned to objective function f_i is a static value which does not change during the algorithm.
- Dynamic sum weighted method: In this approach, the value w_i assigned to objective function f_i is a dynamic value, at each iteration t, we calculate the weight of the iteration $t + 1$, the weights w_i assigned the objective function $f_i(x_t)$ will be updated using the following formula:

$$w_i(t+1) = \frac{\sum_{j=1, j \neq i}^{I} |f^j(x_t)|}{(I-1)\sum_{j=1}^{I} |f^j(x_t)|}, i = 1, ..., m; \tag{4}$$

  For the first iteration ($t = 0$), the weights w_i will be generated randomly, for the remaining of the algorithm, w_i will be updated only if there is an improvement compared to the previous iteration.

3 Proposed Approaches

SA is metaheuristic method inspired from metallurgy that was originally presented in [10], SA is largely used to solve many mono-objective and multiple-objective optimization problems [2, 12, 16], the Metropolis rule helps the algorithm to escape local optima, this is achieved by accepting worse solutions with a certain probability.

Our SA method starts with an initial solution generated by the well-known Prim algorithm, After that, at each iteration, we select randomly a solution x_2 from the neighborhood if the current solution x_1, and which is evaluated using the multiple-objective function Z, the neighbor solution is accepted if its fitness is better that the fitness of the current solution, the neighbor solution is also accepted with a certain probability even if it has a bad fitness, the probability is calculated using the Boltzmann method $P = exp(-Z/T)$, where $Z = f(x_1) - f(x_2)$, and T is the temperature, we decrease the temperature periodically (level of temperature) and not at each iterations, we opted for the geometric cooling schedule, the temperature decreases as follows: $T = \alpha * T$ (with $\alpha < 1$), α is the cooling factor, we note by $RANGE$ the number of iterations performed at each temperature level.

In order to improve the SA method, we have added a restarting strategy to the standard algorithm, In the literature, we find several implementations of this strategy [17]. The restart technique that we have adopted is very simple and effective. The algorithm will be restarted only if no improvement during $TIME\_TO\_RESTART$ secondes, when reaching the restart time, the temperature T will be updated with value recorded when the best solution was found, and also we replace the current solution by the best solution.

The first approach that we denote Dynamic SA, in which we use the dynamic sum weighted sum method is outlined as follows:

- Step 1: Initialization
  - Set α, $RANGE$, $TIME\_TO\_RESTART$
  - Set T_0: the initial temperature
  - Generate randomly the weights w_i for the first iteration
  - Generate randomly an initial solution
- Step 2:
  Repeat until $TIME\_LIMIT$
  - Repeat $RANGE$ times the following instructions:

    Randomly select a neighbor solution.
    Calculate the fitness of the current solution Z_1.
    Calculate the fitness of the selected neighbor Z_2.
    Calculate $Z = Z_2 - Z_1$.
    If $Z < 0$ then the current solution is replaced the neighbor solution and the wights w_i are updated using the formula (4);
    Otherwise, the neighbor solution becomes the current solution and the wights w_i are updated using the formula (4) with a probability equal to exp(−Z/T).
  - If $TIME\_TO\_RESTART$ is reached then restart the algorithm as described before, otherwise the current temperature is decreased.

In the second approach, that we denote Classical SA, we assign a static values to each weight w_i, these values will not be updated during the algorithm, so the complete

Classical SA differs from the Dynamic SA approach only in the way we deal with the weights w_i.

The neighborhood structure adopted to choose the current solution neighbor a is very simple, we randomly delete an edge from the current solution tree, then we randomly choose a vertex from each subtree, and we link them in order to have the neighbor solution tree.

4 Numerical Experiments

In order to compare the performance of each method, we have applied the proposed approaches to solve a problem with two objectives, each edge in the graph has two costs, experiments are performed on a graph that was built by combining two graphs from the library KCTLIB [3](graph 1 $= bb45 \times 5_1.gg$, graph 2 $= bb45 \times 5_2.gg$), both graphs have the same number of vertices ($|V| = 225$), the same number of edges ($|E| = 400$) and also the edge cost values are in [0, 100]. but in order to have a conflicting objectives, we have multiplied each edge costs by 100 in the second graph. The programs are coded in *C* programming language on a MacBook Pro with a processor 1, 4 GHz Intel Core *i*5 4 Core, and memory 8 Go 2133 MHz LPDDR3, we run our programs ten times, then we note the following values:

- Z: The best value obtained for the multi-objective function
- $w_1 f_1$: where w_1 is the weight of the first objective function f_1
- $w_2 f_2$: where w_2 is the weight of the second objective function f_2.

We cannot escape from an empirical adjustment of the SA parameters, the adjustment adopted is as follows:

- $T_0 = 10$
- $\alpha = 0.9$
- $RANGE = 40000$
- $TIME\_TO\_RESTART = 15\,\text{s}$
- $TIME\_STOP = 300\,\text{s}$.

Concerning the dynamic SA approach, the weight w_1 and w_2 are generated randomly only for the first iteration and they are updated in automate way using the formula 5, However, and concerning the classical SA method, we used the following pair of values ($w_1 = 0.3$; $w_2 = 0.7$) and ($w_1 = 0.7$; $w_2 = 0.3$).

Table 1 shows the results obtained by Classical SA and Dynamic SA, the bold values represent the best values among the compared methods.

The result reveals that:

- The Classical SA provides a better performance regarding the optimization of the first objective function $w_1 f_1$, this performance is obtained only when $w_1 = 0.3$ and $w_2 = 0.7$.
- The Dynamic SA shows a better performance regarding the optimization of the second objective function $w_2 f_2$.

Table 1 Comparison results obtained by Classical SA and Dynamic SA

Cardinality		Dynamic SA	Classical SA	
40	Z	**3162.33**	$w_1 = 0.3$; $w_2 = 0.7$	6651.00
			$w_1 = 0.7$; $w_2 = 0.3$	3947.00
	$w_1 f_1$	750.02	$w_1 = 0.3$; $w_2 = 0.7$	**561.00**
			$w_1 = 0.7$; $w_2 = 0.3$	1148.00
	$w_2 f_2$	**2412.30**	$w_1 = 0.3$; $w_2 = 0.7$	6090.00
			$w_1 = 0.7$; $w_2 = 0.3$	2799.00
80	Z	**5191.17**	$w_1 = 0.3$; $w_2 = 0.7$	19279.10
			$w_1 = 0.7$; $w_2 = 0.3$	9946.20
	$w_1 f_1$	2033.60	$w_1 = 0.3$; $w_2 = 0.7$	**1205.10**
			$w_1 = 0.7$; $w_2 = 0.3$	2461.20
	$w_2 f_2$	**3157.57**	$w_1 = 0.3$; $w_2 = 0.7$	18074.00
			$w_1 = 0.7$; $w_2 = 0.3$	7485.00
120	Z	**6680.98**	$w_1 = 0.3$; $w_2 = 0.7$	32945.50
			$w_1 = 0.7$; $w_2 = 0.3$	16769.90
	$w_1 f_1$	3120.08	$w_1 = 0.3$; $w_2 = 0.7$	**1753.50**
			$w_1 = 0.7$; $w_2 = 0.3$	3443.30
	$w_2 f_2$	**3560.90**	$w_1 = 0.3$; $w_2 = 0.7$	31192.00
			$w_1 = 0.7$; $w_2 = 0.3$	13326.60
160	Z	**10168.02**	$w_1 = 0.3$; $w_2 = 0.7$	45098.80
			$w_1 = 0.7$; $w_2 = 0.3$	22153.90
	$w_1 f_1$	4042.85	$w_1 = 0.3$; $w_2 = 0.7$	**2432.40**
			$w_1 = 0.7$; $w_2 = 0.3$	5326.30
	$w_2 f_2$	**6125.17**	$w_1 = 0.3$; $w_2 = 0.7$	42666.40
			$w_1 = 0.7$; $w_2 = 0.3$	16827.60
200	Z	**15814.61**	$w_1 = 0.3$; $w_2 = 0.7$	58856.60
			$w_1 = 0.7$; $w_2 = 0.3$	31151.10
	$w_1 f_1$	5139.40	$w_1 = 0.3$; $w_2 = 0.7$	**2974.20**
			$w_1 = 0.7$; $w_2 = 0.3$	6218.10
	$w_2 f_2$	**10675.21**	$w_1 = 0.3$; $w_2 = 0.7$	55882.40
			$w_1 = 0.7$; $w_2 = 0.3$	24933.00

- The Dynamic SA shows a better performance regarding the optimization of the multiple objective function z.

If we take in consideration that the edges cost of the second objective are 10 times bigger than the edge costs of the first objective, we can say that the best method is the one which provides better results regarding the second objective, and therefore, we can conclude that the Dynamic SA approach is better than the Classical SA, this conclusion is also confirmed by the results obtained by the Dynamic SA regarding the optimization of the multiple-objective function Z. The good results of Dynamic

SA are obtained thanks to the dynamic weighted sum method, this method ensure an equitable treatment of all objectives by automating the calculation of their weights w_i during the optimization process.

5 Conclusion

In this work, we have tried to provide two approaches based on SA algorithm in order to solve the MO k-MST problem. First, the multi-objective optimization problem is transformed into a mono-objective one by using two weighted sum methods, namely the dynamic weighted sum method and the classical weighted sum method. In order to compare the performance of the proposed approaches, we have conducted some numerical experiments, their results showed that the dynamic SA method is better than the classical SA. In future works, we will develop some approximate approches based on the dynamic weighted sum method in order to solve the MO k-MST in case of fuzzy problems.

References

1. Arroyo, J.E., Vieira, P., Vianna, D.: A GRASP algorithm for the multi-criteria minimum spanning tree problem. Ann. OR **159**, 125–133 (2008)
2. Baños, R., Ortega, J., Gil, C., Fernández, A., de Toro, F.: A simulated annealing-based parallel multi-objective approach to vehicle routing problems with time windows. Exp. Syst. Appl. **40**(5), 1696–1707 (2013)
3. Blum, C., Blesa, M.J.: New metaheuristic approaches for the edge-weighted k-cardinality tree problem. Comput. Oper. Res. **32**(6), 1355–1377 (2005)
4. Chankong, V., Haimes, Y.: Multiobjective Decision Making: Theory and Methodology (1983)
5. Davis-Moradkhan, M., Browne, W.: Evolutionary algorithms for the multi criterion minimum spanning tree problem. In: Tenne, Y., Goh, C.-K. (eds.) Computational Intelligence in Expensive Optimization Problems. Adaptation Learning and Optimization, pp. 423–452. Springer, Berlin, Heidelberg (2010)
6. Gabli, M., Jaara, E., El Bekkaye, M.: A genetic algorithm approach for an equitable treatment of objective functions in multi-objective optimization problems. IAENG Int. J. Comput. Sci. **41**, 102–111 (2014)
7. Goldbarg, E., Souza, G., Goldbarg, M.: Particle Swarm Optimization for the Bi-objective Degree Constrained Minimum Spanning Tree, pp. 420–427, Jan. 2006
8. Guo, W., Chen, G., Feng, X., Yu, L.: Solving Multi-criteria Minimum Spanning Tree Problem with Discrete Particle Swarm Optimization, pp. 471–478, Sept. 2007
9. Han, L., Wang, Y.: A novel genetic algorithm for multi-criteria minimum spanning tree problem. In: Hao, Y., Liu, J., Wang, Y., Cheung, Y.-M., Yin, H., Jiao, L., Ma, J., Jiao, Y.-C. (eds.), Computational Intelligence and Security, Lecture Notes in Computer Science, pp. 297–302. Springer, Berlin, Heidelberg (2005)
10. Kirkpatrick, S., Gelatt, C.D., Vecchi, M.P.: Optimization by simulated annealing. Science (New York, N.Y.), vol. 220(4598), pp. 671–680, May 1983
11. Knowles, J., Corne, D.: A comparison of encodings and algorithms for multiobjective minimum spanning tree problems. In: Proceedings of the IEEE Conference on Evolutionary Computation, ICEC, 1 Apr. 2001

12. Liu, L., Haibo, M., Yang, J., Li, X., Fang, W.: A simulated annealing for multi-criteria optimization problem: DBMOSA. Swarm Evol. Comput. **14**, 48–65 (2014)
13. Liu, Q., Li, X., Liu, H., Guo, Z.X.: Multi-objective metaheuristics for discrete optimization problems: a review of the state-of-the-art. Appl. Soft Comput. **93**, 106382, May 2020
14. Narzisi, G.L.: Classic Methods for Multi-objective Optimization (2008)
15. Neumann, F.: Expected runtimes of a simple evolutionary algorithm for the multi-objective minimum spanning tree problem. European J. Oper. Res. **181**(3), 1620–1629 (2007)
16. Robini, M.C., Reissman, P.-J.: From simulated annealing to stochastic continuation: a new trend in combinatorial optimization. J. Glob. Optim. **56**(1), 185–215 (2013)
17. Yu, V.F., Redi, A.A.N.P., Hidayat, Y.A., Wibowo, O.J.: A simulated annealing heuristic for the hybrid vehicle routing problem. Appl. Soft Comput. **53**, 119–132, Apr. 2017
18. Zhou, G., Gen, M.: Genetic algorithm approach on multi-criteria minimum spanning tree problem. European J. Oper. Res. **114**(1), 141–152 (1999)

Kantorovich Methods for Urysohn Integral Equations

M. Arrai, C. Allouch, and M. Tahrichi

Abstract In this paper, the *Kantorovich* method for the numerical solution of non-linear *Urysohn* equations with a smooth kernel is considered. The approximating operator is chosen to be either the orthogonal projection or an interpolatory projection onto a space of piecewise polynomials of degree $\leq r-1$. This method have asymptotic series expansions and the orders of convergence can be further improved by the *Richardson* extrapolation, assuming the calculation to be repeated with each subinterval halved. We show that these orders of convergence are preserved in the corresponding discrete methods obtained by calculating the integrals with a numerical quadrature formula. Numerical examples are given to illustrate the theoretical estimates.

Keywords *Urysohn* equation · *Kantorovich* method · Projection operator · *Gauss* points · Extrapolation · Discrete methods

1 Introduction

We consider the following *Urysohn* integral equation defined on $\mathbb{X} = \mathscr{L}^{\infty}[0, 1]$ by

$$x(s) - \int_0^1 \kappa(s, t, x(t))dt = f(s), \quad s \in [0, 1] \tag{1}$$

M. Arrai (✉) · C. Allouch
FPN, MSC Team, LAMAO Laboratory, University Mohammed I, Nador, Morocco
e-mail: mohamedarrai28@gmail.com

C. Allouch
e-mail: c.allouch@ump.ma

M. Tahrichi
ESTO, ANAA Team, ANO Laboratory, University Mohammed I, Oujda, Morocco
e-mail: mtahrichi@hotmail.com

S. Melliani et al. (eds.), *Applied Mathematics and Modelling in Finance, Marketing and Economics*, Studies in Computational Intelligence 1114,
https://doi.org/10.1007/978-3-031-42847-0_5

where the kernel $\kappa(s, t, u)$ is a real smooth function and x is the unknown function to be determined.

Classical methods for solving (1) are the *Galerkin* method based on the orthogonal projection onto a finite dimensional subspace of $\mathbb{X}$ and the collocation method based on an interpolatory projection. The iterated *Galerkin*/iterated collocation solutions are obtained by one step of iteration and were studied for *Urysohn* integral equations in [4]. The discrete version of collocation/iterated collocation methods was considered in *Atkinson-Flores* [3]. Recently, a different method, called modified projection method, was introduced in [6], while its discrete version was proposed in *Kulkarni-Rakshit* [10]. The obtained solution is shown to converge faster than the iterated *Galerkin* solution. More recently, superconvergent *Nyström* method was used in [2] to solve Eq. (1) with smooth kernels which converges as rapid as the modified projection method.

Asymptotic error analysis is a classical numerical analysis topic for improving the orders of convergence of the approximate solutions. If the error expansions for numerical solutions are established, then the *Richardson* extrapolation can then be used to obtain approximate solutions of higher order. For nonlinear integral equations, *Guoqiang* [7] obtained asymptotic error expansion for the discrete *Kumar* and *Sloan* solution, and for the iterated projection and iterated modified projection solution, was proved by *Kulkarni* and *Nidhin* [11]. The problem of asymptotic expansion for an approximate solution of a nonlinear *Hammerstein* equation, was considered in [1]. It is considered in the case of a superconvergent projection-type method using the orthogonal projection or an interpolatory projection. The asymptotic series expansion for the discrete iterated modified projection solution was discussed by *Kulkarni-Rakshit* in [9].

The purpose of this paper is to investigate the *Kantorovich* method for solving (1), which is based on "Kantorovich regularization" (*Kantorovich*, 1948) using piecewise polynomial basis functions. This method is discussed in *Schock* [14] and *Sloan* [15] for *Fredholm* integral equations and it seems not to be studied at the moment for nonlinear integral equations. If the right hand side of the operator equation is less smooth than the kernel of the integral operator, then the *Kantorovich* solution has a higher order of convergence than the *Galerkin* solution. We define also the iterated *Kantorovich* method and we establish that it had a faster convergence than the *Kantorovich* method. Moreover, we give an asymptotic error expansion of the iterated *Kantorovich* method for (1). Thus, *Richardson* extrapolation can be performed on the solution, and this will increase greatly the accuracy of numerical solution. We show that the obtained orders of convergence are still valid after taking into account the errors introduced by the numerical quadrature formula.

Now for a summary of the paper. In Sect. 2, notation is set, the numerical methods are described, and some relevant results are recalled. In Sect. 3, asymptotic series expansion for the iterated *Kantorovich* method with both the orthogonal projection and the interpolatory projection at *Gauss* points is obtained. Sect. 4. is devoted to the discrete version of the proposed methods. In Sect. 5, we illustrate our results by numerical examples.

2 Preliminaries and Method

For a positive integer n, let

$$(\Delta_n): \quad 0 = s_0 < s_1 < s_2 < \cdots < s_{n-1} < s_n = 1 \tag{2}$$

be the uniform partition of $[0, 1]$, with nodes $\{s_i = \frac{i}{n}, \ i = 0, \ldots, n\}$ and meshlength $h = \frac{1}{n}$. For a fixed $r \geq 1$, we denote by Π_r the space of polynomials of degree $\leq r - 1$. Let

$$\mathbb{X}_n = \left\{ y : [0, 1] \longrightarrow \mathbb{R} : y|_{[s_{i-1}, s_i]} \in \Pi_r, \ 1 \leq i \leq n \right\}$$

be the set of functions that are polynomials of degree $\leq r - 1$, on each subinterval $[s_{i-1}, s_i]$. We use two types of projections from $\mathbb{X}$ to $\mathbb{X}_n$.

• The map π_n is the restriction to $\mathbb{X}$ of the orthogonal projection from $\mathscr{L}^2[0, 1]$ to $\mathbb{X}_n$. Then

$$(\pi_n x)(s) = \sum_{i=0}^{nr} \langle x, \varphi_i \rangle \varphi_i(s), \tag{3}$$

where $\{\varphi_1, \varphi_2, \ldots, \varphi_{nr}\}$ is an orthonormal basis for $\mathbb{X}_n$ and $\langle ., . \rangle$ is the inner product in $\mathscr{L}^2[0, 1]$.

• For $x \in C[0, 1]$, let $\pi_n x$ denote the unique piecewise polynomial of degree $r - 1$ that satisfies

$$(\pi_n x)(t_{ij}) = x(t_{ij}), \tag{4}$$

where the collocation points are

$$t_{ij} = (i - 1 + \tau_j)h, \quad 1 \leq i \leq n, \quad 1 \leq j \leq r, \tag{5}$$

and $\{\tau_1, \ldots, \tau_r\}$ are the r *Gauss* points in $[0, 1]$. This map, if necessary, is extended to $\mathbb{X}$ and then π_n is a projection. In both cases, π_n converge to identity operator pointwise and for $x \in C^r[0, 1]$,

$$\|x - \pi_n x\|_\infty \leq c_1 \|x^{(r)}\|_\infty h^r, \tag{6}$$

where c_1 is a constant independent of n. Moreover, the projection π_n is uniformly bounded with respect to n, i.e.

$$P = \sup_n \|\pi_n\|_{\mathbb{X} \to \mathbb{X}} < \infty. \tag{7}$$

Let $x, y \in C^r[0, 1]$. If π_n is the restriction of the orthogonal projection to $\mathscr{L}^\infty[0, 1]$, then it follows from (6) that

$$\left|\int_0^1 x(t)(I-\pi_n)y(t)dt\right| = |\langle (I-\pi_n)x, (I-\pi_n)y\rangle| \le (c_1)^2\|x^{(r)}\|_\infty \|y^{(r)}\|_\infty h^{2r}. \tag{8}$$

Let p be a positive integer. For $x \in C^p[0, 1]$, we set

$$\|x\|_{p,\infty} = \sum_{i=0}^{p} \|x^{(i)}\|_\infty.$$

If π_n is the interpolatory projection at r *Gauss* points, then for $x \in C^r[0, 1]$ and $y \in C^{2r}[0, 1]$, (See *de-Boor-Swartz* [5]),

$$\left|\int_0^1 x(t)(I-\pi_n)y(t)dt\right| \le c_2\|x\|_{r,\infty}\|y\|_{2r,\infty}h^{2r}, \tag{9}$$

where c_2 is a constant independent of n.

Let $\mathcal{K}$ be the *Urysohn* integral operator defined by

$$(\mathcal{K}x)(s) = \int_0^1 \kappa(s, t, x(t))dt, \quad s \in [0, 1]. \tag{10}$$

Thus, Eq. (1) can be writing in operator form as

$$x - \mathcal{K}(x) = f. \tag{11}$$

For our convenience we let

$$z = \mathcal{K}(x). \tag{12}$$

Thus, writing the solution of (11) as $x = z + f$, we have

$$z = \mathcal{K}(z + f). \tag{13}$$

The *Kantorovich* method, is obtained by applying the classical projection method to the Eq. (13). Thus, the approximate solution is

$$x_n = z_n + f, \tag{14}$$

where z_n satisfies

$$z_n - \pi_n\mathcal{K}(z_n + f) = 0. \tag{15}$$

The theoretical advantage of the proposed method is that the inhomogeneous term is now 0 rather than $\pi_n f$ in projection methods which may be smoother than f. Note that the above equations are equivalent to a single equation for x_n

$$x_n - \pi_n \mathscr{K}(x_n) = f. \tag{16}$$

Throughout this paper, this method will be called respectively a *Kantorovich-Galerkin* or *Kantorovich*-collocation method when the orthogonal projection or the interpolatory projection is used.

Finally, the iterated *Kantorovich* approximation is defined by

$$\begin{aligned} \widetilde{x}_n &= \mathscr{K}(x_n) + f, \\ &= \widetilde{z}_n + f, \end{aligned} \tag{17}$$

where

$$\widetilde{z}_n = \mathscr{K}(z_n + f). \tag{18}$$

From (15) and (17) we observe that $z_n = \pi_n \widetilde{z}_n$, and hence

$$\widetilde{z}_n - \mathscr{K}(\pi_n \widetilde{z}_n + f) = 0. \tag{19}$$

For the implementation of the method, we define

$$F_n(y) = y - \pi_n \mathscr{K}(y + f).$$

Then, Eq. (15) becomes

$$F_n(z_n) = 0.$$

This last equation is solved iteratively by using the *Newton-Kantorovich* method. For an initial approximation $z_n^{(0)}$, define

$$z_n^{(k+1)} = z_n^{(k)} - [F_n'(z_n^{(k)})]^{-1} F_n(z_n^{(k)}),$$

where $F_n'(z_n^{(k)})$ is the *Fréchet* derivative of F_n given by

$$F_n'(z_n^{(k)})h = h - \pi_n \mathscr{K}'(z_n^{(k)} + f)h.$$

By a simple calculus, we get

$$z_n^{(k+1)} - \pi_n \mathscr{K}'(z_n^{(k)}) z_n^{(k+1)} = \pi_n \mathscr{K}(z_n^{(k)} + f) - \pi_n \mathscr{K}'(z_n^{(k)}) z_n^{(k)}. \tag{20}$$

Since $z_n^{(k)} \in \mathbb{X}_n$, we can write in the case of orthogonal projection

$$z_n^{(k)} = \sum_{j=1}^{nr} \langle z_n^{(k)}, \varphi_j \rangle \varphi_j = \sum_{j=1}^{nr} y_n^{(k)}(j) \varphi_j.$$

Then, (20) is equivalent to the following linear system of size nr

$$(I - A_n^{(k)})y_n^{(k+1)} = r_n^{(k)},$$

where for $i, j = 1, \ldots, nr$,

$$\begin{aligned} A_n^{(k)}(i,j) &= \langle \mathcal{K}'(z_n^{(k)})\varphi_j, \varphi_i \rangle, \\ r_n^{(k)}(i) &= \langle \mathcal{K}(z_n^{(k)} + f), \varphi_i \rangle - (C_n^{(k)} y_n^{(k)})(i). \end{aligned}$$

Let $\{\ell_1, \ldots, \ell_{nr}\}$ be the *Lagrange* basis of $\mathbb{X}_n$ satisfying

$$\ell_i(t_j) = \begin{cases} 1 \text{ if } & i = j, \\ 0 \text{ if } & i \neq j, \end{cases}$$

where $\{t_1, \ldots, t_{nr}\}$ are the ordered interpolation points given by (5). For the interpolatory projection, we can write

$$z_n^{(k)} = \sum_{j=1}^{nr} z_n^{(k)}(t_j)\ell_j = \sum_{j=1}^{nr} y_n^{(k)}(j)\ell_j.$$

Then, we obtain the system of linear equations

$$(I - B_n^{(k)})y_n^{(k+1)} = q_n^{(k)},$$

where for $i, j = 1, \ldots, nr$,

$$\begin{aligned} B_n^{(k)}(i,j) &= \mathcal{K}'(z_n^{(k)})(t_j), \\ q_n^{(k)} &= \mathcal{K}(z_n^{(k)} + f)(t_i) - (B_n^{(k)} y_n^{(k)})(i). \end{aligned}$$

3 Convergence Rates

Let x_0 be an isolated solution of (1), and let a, b be real numbers such that

$$\left[\min_{s\in[0,1]} x_0(s), \max_{s\in[0,1]} x_0(s)\right] \subset [a, b].$$

Define

$$\Omega = [0,1] \times [0,1] \times [a,b].$$

Let $\alpha \geq 1$. For the rest of this section, we assume that

$$\kappa \in C^{\alpha}(\Omega) \quad \text{and} \quad \frac{\partial \kappa}{\partial u} \in C^{2\alpha}(\Omega).$$

Then, $\mathcal{K}$ is a compact operator from $\mathcal{L}^{\infty}[0, 1]$ to $C^{\alpha}[0, 1]$. If $f \in C[0, 1]$, then, since

$$x_0 - \mathcal{K}(x_0) = f, \tag{21}$$

the solution x_0 belongs to $C[0, 1]$. Moreover, the operator $\mathcal{K}$ is *Fréchet* differentiable and the *Fréchet* derivative is given by

$$(\mathcal{K}'(x)g)(t) = \int_0^1 \frac{\partial \kappa}{\partial u}(s, t, x(t))g(t)dt.$$

Also, the second derivative $\mathcal{K}''(x)$ is the bi-linear function given by

$$(\mathcal{K}''(x)(g_1, g_2))(s) = \int_0^1 \frac{\partial^2 \kappa}{\partial u^2}(s, t, x(t))g_1(t)g_2(t)dt.$$

For $\delta_0 > 0$, let

$$\mathcal{B}(x, \delta_0) = \{y \in \mathbb{X} : \|x - y\|_{\infty} < \delta_0\}.$$

Since $\frac{\partial \kappa}{\partial u} \in C^{2\alpha}(\Omega)$, it follows that $\mathcal{K}'$ is *Lipschitz* continuous in a neighborhood $\mathcal{B}(x_0, \delta_0)$ of x_0, that is, there exists a constant γ such that

$$\|\mathcal{K}'(x_0) - \mathcal{K}'(x)\| \leq \gamma \|x_0 - x\|, \quad x \in \mathcal{B}(x, \delta_0). \tag{22}$$

The operator $\mathcal{K}'(x_0)$ is compact. Assume that $(I - \mathcal{K}'(x_0))^{-1} : C[0, 1] \to C[0, 1]$ is a bounded linear operator and that 1 is not an eigenvalue of $\mathcal{K}'(x_0)$. Then it can be shown that

$$M = (I - \mathcal{K}'(x_0))^{-1}\mathcal{K}'(x_0)$$

is the compact linear integral operator (see [12]) given by

$$(Mg)(s) = \int_0^1 m(s, t)g(t)dt,$$

where the smoothness of kernel m is the same as that of kernel

$$\ell(s, t) = \frac{\partial \kappa}{\partial u}(s, t, x_0),$$

that is,

$$m \in C^{2\alpha}([0, 1] \times [0, 1]).$$

The following lemma, which can be shown easily, will be used to prove the main results of this section.

Lemma 1 *Suppose that $x_0 \in C[0,1]$ is an isolated solution of (1) and assume that 1 is not an eigenvalue of $\mathcal{K}'(x_0)$. Then for n large enough, the operators $I - (\pi_n \mathcal{K})'(x_0)$ are invertible i.e. there exists a constant $A > 0$ such that*

$$\|(I - (\pi_n \mathcal{K})'(x_0))^{-1}\|_\infty \leq A < \infty.$$

The following theorem can be proved by using Theorem 2 of *Vainikko* [16].

Theorem 1 *Suppose that $x_0 \in C[0,1]$ is an isolated solution of (1) and that 1 is not an eigenvalue of $\mathcal{K}'(x_0)$. Then there exists a real number $\delta_0 > 0$ such that the approximate equation (16) has a unique solution x_n in $\mathcal{B}(x_0, \delta_0)$ for a sufficiently large n. Moreover, there exists a constant $0 < q < 1$, independent of n such that*

$$\frac{\alpha_n}{1+q} \leq \|x_n - x_0\|_\infty \leq \frac{\alpha_n}{1-q}, \tag{23}$$

where $\alpha_n = \left\|(I - (\pi_n \mathcal{K})'(x_0))^{-1}(\mathcal{K}(x_0) - \pi_n \mathcal{K}(x_0))\right\|_\infty \longrightarrow 0$ *as* $n \longrightarrow \infty$.

The next theorem establish the rate of convergence of the approximation x_n to the exact solution x_0.

Theorem 2 *Let x_0, x_n be the solutions of (11) and (16) respectively. Assume that $\alpha \geq r$. Then, under the hypothesis of Theorem 1, for n large enough, we have*

$$\|x_n - x_0\|_\infty = O(h^r). \tag{24}$$

Proof The result is a direct consequence of estimates (6), (23) and Lemma 1.

An enhancement in the rate of convergence is obtained in the following theorem.

Theorem 3 *Let π_n be either the restriction to $\mathcal{L}^\infty[0,1]$ of the orthogonal projection from $\mathcal{L}^2[0,1]$ to $\mathbb{X}_n$ or the interpolatory projection at r* Gauss *points in each subinterval of the partition. Assume that $\alpha \geq r+1$ and that 1 is not an eigenvalue of $\mathcal{K}'(x_0)$. Then*

$$\|\widetilde{x}_n - x_0\| = O(h^{2r}). \tag{25}$$

Proof Note that from Eqs. (17) and (21) we have

$$\begin{aligned} \widetilde{x}_n - x_0 &= \mathcal{K}(x_n) - \mathcal{K}(x_0) \\ &= \mathcal{K}(x_n) - \mathcal{K}'(x_0)(x_n - x_0) + \mathcal{K}'(x_0)(x_n - x_0) - \mathcal{K}(x_0). \end{aligned} \tag{26}$$

Noting that

$$x_n - x_0 = \pi_n(\widetilde{x}_n - x_0) - (I - \pi_n)\mathcal{K}(x_0), \tag{27}$$

yields

$$\begin{aligned} \widetilde{x}_n - x_0 &= [\mathcal{K}(x_n) - \mathcal{K}'(x_0)(x_n - x_0) - \mathcal{K}(x_0)] \\ &\quad + \mathcal{K}'(x_0)\pi_n(\widetilde{x}_n - x_0) - \mathcal{K}'(x_0)(I - \pi_n)\mathcal{K}(x_0). \end{aligned} \tag{28}$$

Hence, using again (26), we get

$$\mathscr{K}'(x_0)\pi_n(\widetilde{x}_n - x_0) = \mathscr{K}'(x_0)(\pi_n - I)[\mathscr{K}(x_n) - \mathscr{K}'(x_0)(x_n - x_0) + \mathscr{K}'(x_0)(x_n - x_0) - \mathscr{K}(x_0)] + \mathscr{K}'(x_0)(\widetilde{x}_n - x_0)$$

and replacing in (28), we obtain the formula

$$\begin{aligned} \widetilde{x}_n - x_0 = \{ & \left[I - \mathscr{K}'(x_0)\right]^{-1} [\mathscr{K}(x_n) - \mathscr{K}'(x_0)(x_n - x_0) - \mathscr{K}(x_0)] \\ & - M(I - \pi_n)[\mathscr{K}(x_n) - \mathscr{K}'(x_0)(x_n - x_0) - \mathscr{K}(x_0)] \\ & - M(I - \pi_n)\mathscr{K}'(x_0)(x_n - x_0) - M(I - \pi_n)\mathscr{K}(x_0)\}. \end{aligned} \tag{29}$$

By the mean value theorem for $0 < \theta < 1$, and using the *Lipschitz* continuity of $\mathscr{K}'$, we obtain

$$\begin{aligned} \|\mathscr{K}(x_n) - \mathscr{K}'(x_0)(x_n - x_0) - \mathscr{K}(x_0)\| &= \|[\mathscr{K}'(x_n + \theta(x_0 - x_n)) - \mathscr{K}'(x_0)](x_n - x_0)\| \\ &\leq \gamma(1 - \theta)\|x_n - x_0\|_\infty^2. \end{aligned} \tag{30}$$

Using (8) and (9), we can show that

$$\|M(I - \pi_n)\mathscr{K}(x_0)\| = \mathscr{O}(h^{2r}) \tag{31}$$

and in [8, Lemma 2.1], it is shown that

$$\|M(I - \pi_n)\mathscr{K}'(x_0)\| = \mathscr{O}(h^{2r}). \tag{32}$$

Thus, combining the estimates (29)–(32), the result follows.

Let $B_0(t) = 1$ and for $i \geq 1$, let $B_i(\tau)$ denote the *Bernoulli* polynomial of degree i. Let $\eta_1, \eta_2, \ldots,$ be the sequence of orthonormal polynomials in $\mathscr{L}^2[0, 1]$ i.e. η_p is a polynomial of degree $p - 1$, and

$$\langle \eta_p, \eta_q \rangle = \delta_{pq} \quad \text{for all } p, q \geq 1.$$

Define

$$\Lambda_r(\sigma, \tau) = \sum_{p=1}^{r} \eta_p(\sigma)\eta_p(\tau)$$

and

$$\chi_j(\tau) = \int_0^1 \Lambda_r(\sigma, \tau)\frac{(\sigma - \tau)^j}{j!}d\sigma, \quad j = 1, \ldots, r + 1.$$

Let π_n be the orthogonal projection from $\mathscr{L}^2[0, 1]$ to $\mathbb{X}_n$ and assume that $g \in C^{2r+2}[0, 1]$. Let T denote the linear integral operator with kernel $q(.,.) \in C^{2r+2}[0, 1]^2$ defined by

$$(Tu)(s) = \int_0^1 q(s,t)u(t)dt, \quad u \in \mathbb{X}, \quad s \in [0,1].$$

Then, the following asymptotic series expansion is proved in [13]

$$T(I - \pi_n)g = \mathscr{U}(g)h^{2r} + \mathscr{O}(h^{2r+2}), \tag{33}$$

where for the orthogonal projection

$$\mathscr{U}(g)(s) = a_{2r}(Tg^{(2r)})(s) + \sum_{i=1}^{2r-1} a_i \left[\left(\frac{\partial}{\partial t}\right)^{2r-1-i} q(s,t)g^{(i)}(t) \right]_{t=0}^{1},$$

and

$$a_i = \int_0^1 \int_0^1 \Lambda_r(\sigma,\tau) \frac{B_{2r-i}(\tau)}{(2r-i)!} \frac{(\sigma-\tau)^i}{i!} d\sigma d\tau, \quad i = 1, \ldots, 2r,$$

while for the interpolatory projection

$$\mathscr{U}(g)(s) = b_{2r}(Tg^{(2r)})(s) + \sum_{i=r}^{2r-1} b_i \left[\left(\frac{\partial}{\partial t}\right)^{2r-1-i} q(s,t)g^{(i)}(t) \right]_{t=0}^{1}, \tag{34}$$

with

$$b_i = -\int_0^1 \Phi_i(\tau) \frac{B_{2r-i}(\tau)}{(2r-i)!} \Psi_r(\tau) d\tau, \tag{35}$$

$$\Phi_i(\tau) = \int_0^1 \frac{(\sigma-\tau)^{i-r}}{(i-r)!} \frac{[\tau_1, \ldots, \tau_r, \tau](\bullet - \sigma)_+^{r-1}}{(r-1)!} d\sigma, \tag{36}$$

and

$$\Psi_r(\tau) = \prod_{i=1}^{r} (\tau - \tau_i).$$

The following asymptotic expansion for the iterated *Kantorovich* solution $\widetilde{x}_n$ can be proved by using technics from [11].

Theorem 4 *Let π_n be either the restriction to $\mathscr{L}^\infty[0,1]$ of the orthogonal projection from $\mathscr{L}^2[0,1]$ to $\mathbb{X}_n$ or the interpolatory projection at r Gauss points in each subinterval of the partition. Assume that $\alpha \geq r+1$ and that 1 is not an eigenvalue of $\mathscr{K}'(x_0)$. Then*

$$\widetilde{x}_n = x_0 + \eta h^{2r} + \mathscr{O}(h^{2r+2}), \tag{37}$$

where

$$\eta = \frac{1}{2}\left[I - \mathscr{K}'(x_0)\right]^{-1} V(x_0) + U(x_0)$$

and for the orthogonal projection

$$V(x) = \left(\int_0^1 \chi_r(\tau)^2 d\tau \right) \mathscr{K}''(x_0) \left(x^{(r)} \right)^2$$

while for the interpolatory projection

$$V(x) = \left(\int_0^1 \Psi_r(\tau)^2 \Phi_r(\tau)^2 d\tau \right) \mathscr{K}''(x_0) \left(x^{(r)} \right)^2.$$

One step of *Richardson* extrapolation can be used to further improve the order of convergence of $\widetilde{x}_n$. Define

$$x_n^R = \frac{2^{2r} \widetilde{x}_{2n} - \widetilde{x}_n}{2^{2r} - 1}.$$

Then since

$$\widetilde{x}_n = x_0 + \eta h^{2r} + \mathscr{O}(h^{2r+2}),$$

$$\widetilde{x}_{2n} = x_0 + \eta \left(\frac{h}{2} \right)^{2r} + \mathscr{O}(h^{2r+2}),$$

we have

$$x_n^R = x_0 + \mathscr{O}(h^{2r+2}). \tag{38}$$

4 Discrete Methods

In practice, the integrals in the definitions of the orthogonal projection π_n and the operator $\mathscr{K}$ involved in Eqs. (3) and (10) are not computed exactly. It is necessary to replace them by a numerical quadrature formula giving rise to discrete methods. In this section, we investigate the discrete version of *Kantorovich*-collocation method and the analysis can be extended to *Kantorovich-Galerkin* method. We consider a basic quadrature formula defined by

$$\int_0^1 f(t) dt \simeq \sum_{j=1}^{R} w_j f(\sigma_j), \tag{39}$$

with nodes $\sigma_1, \sigma_2, \ldots, \sigma_R \in [0, 1]$ and weights are such that

$$\sum_{j=1}^{R} w_j = 1.$$

Let $m \in \mathbb{N}$ and let Δ_m be the uniform partition of [0, 1] giving by (2) with meshlength $\tilde{h} = \frac{1}{m}$. For $1 \leq i \leq m$ and $1 \leq j \leq R$, let $\zeta_{ij} = (i - 1 + \sigma_j)\tilde{h}$, then (39) gives rise to the composite quadrature formula

$$\int_0^1 f(t)dt \simeq \tilde{h} \sum_{i=1}^{m} \sum_{j=1}^{R} w_j f(\zeta_{ij}). \tag{40}$$

Suppose that the quadrature formula (39) is exact for all polynomials of degree $\leq d - 1$. Then, for $f \in C^d[0, 1]$

$$\left| \int_0^1 f(t)dt \simeq \tilde{h} \sum_{i=1}^{m} \sum_{j=1}^{R} w_j f(\zeta_{ij}) \right| \leq c_1 \|f^{(d)}\|_\infty \tilde{h}^d, \tag{41}$$

where c_1 is a constant independent of m.
The *Nyström* approximation of the integral operator $\mathcal{K}$ is defined as

$$(\mathcal{K}_m x)(s) = \tilde{h} \sum_{i=1}^{m} \sum_{j=1}^{R} w_j \kappa(s, \zeta_{ij}, x(\zeta_{ij})), \quad s \in [0, 1]. \tag{42}$$

Assume that $\kappa \in C^d(\Omega)$ and that $x \in C^d[0, 1]$. Then

$$\|\mathcal{K}(x) - \mathcal{K}_m(x)\|_\infty = O(\tilde{h}^d). \tag{43}$$

The *Fréchet* derivative of $\mathcal{K}_m$ is given by

$$(\mathcal{K}'_m(x)g)(s) = \tilde{h} \sum_{i=1}^{m} \sum_{j=1}^{R} w_j \frac{\partial \kappa}{\partial u}(s, \zeta_{ij}, x(\zeta_{ij}))g(\zeta_{ij}), \quad s \in [0, 1]$$

If $\frac{\partial \kappa}{\partial u} \in C^d(\Omega)$ and $g \in C^d[0, 1]$, then from (41)

$$\|\mathcal{K}'(x_0)g - \mathcal{K}'_m(x_0)g\| \leq c_2 \|g\|_{d,\infty} \tilde{h}^d, \tag{44}$$

where c_2 is a constant independent of m.
Let π_n be the interpolatory projection given by (4). Replacing the operator $\mathcal{K}$ with $\mathcal{K}_m$ in (16), we obtain the discrete *Kantorovich*-collocation method

$$y_n - \pi_n \mathcal{K}_m(y_n) = f. \tag{45}$$

We can show that for n and m large enough, the above equation has a unique solution in a neighbourhood $\mathcal{B}(x_0, \delta)$ of x_0, where $\delta > 0$.
The discrete iterated *Kantorovich* solution is defined as follows :

$$\widetilde{y}_n = \mathscr{K}_m(y_n) + f. \tag{46}$$

Proposition 1 *([10]) If* $\frac{\partial \kappa}{\partial u} \in C^r(\Omega)$ *and* $g \in C^{2r}[0, 1]$, *then*

$$\|\mathscr{K}_m'(x_0)(I - \pi_n)g\|_\infty \leq c_3 \|\ell\|_{r,\infty} \|g\|_{2r,\infty} h^{2r},$$

where $c_3 = \frac{2r}{r!}\|\Psi\|_\infty \left(\sum_{j=1}^{R} |\omega_j|\right)$ *is a constant independent of* n.

In the rest of the paper, we choose $d \geq r$ and $m = pn$ for some $p \in \mathbb{N}^*$. Then $\tilde{h} = \frac{h}{p}$.

Theorem 5 *Let* x_0, y_n *be the solutions of (11) and (45) respectively. Assume that* $\alpha \geq r$. *Then, for a sufficiently large* n, *we have*

$$\|x_0 - y_n\|_\infty = O(h^r). \tag{47}$$

Proof Again by Theorem 2 of *Vainikko* [16], we can show that

$$\|x_0 - y_n\|_\infty \leq c \big\| (I - (\pi_n \mathscr{K}_m)'(x_0))^{-1} \| \| (\mathscr{K}(x_0) - \pi_n \mathscr{K}_m(x_0)) \big\|_\infty, \tag{48}$$

where c is a constant independent of n. Since $(I - (\pi_n \mathscr{K}_m)'(x_0))^{-1}$ exists, there is a $m_1 \in \mathbb{N}$ such that for $m \geq m_1$

$$\|(I - (\pi_n \mathscr{K}_m)'(x_0))^{-1}\| \leq 2\|(I - (\pi_n \mathscr{K})'(x_0))^{-1}\| \leq 2A. \tag{49}$$

On the other hand, it follows from (6), (7) and (26) that

$$\begin{aligned}
\|(\mathscr{K}(x_0) - \pi_n \mathscr{K}_m(x_0))\|_\infty &\leq \|(I - \pi_n)[\mathscr{K}_m(x_0) - \mathscr{K}(x_0)])\|_\infty \\
&\quad + \|(I - \pi_n)\mathscr{K}(x_0)\| \|(I - \pi_n)\mathscr{K}_m(x_0)\|_\infty \\
&\quad + \|\mathscr{K}(x_0) - mathscrK_m(x_0)\|_\infty, \\
&= O(\max\{h^r, \tilde{h}^d\}).
\end{aligned}$$

This completes the proof.

The following theorem gives the order of convergence in the discrete iterated *Kantorovich* method.

Theorem 6 *Let* $\widetilde{y}_n$ *be the discrete iterated solution defined by (46). Assume that* $\kappa \in C^d(\Omega)$, $\frac{\partial \kappa}{\partial u} \in C^d(\Omega)$ *and that* $f \in C[0, 1]$. *Then, for a sufficiently large* n, *we have*

$$\|x_0 - \widetilde{y}_n\|_\infty = O(h^{2r}). \tag{50}$$

Proof For n big enough it can be easily checked that

$$\|x_0 - \tilde{y}_n\| \leq c\|\mathscr{K}(z_0 + f) - \mathscr{K}_m(\pi_n z_0 + f)\|.$$

We write

$$\begin{aligned}\|\mathscr{K}(z_0 + f) - \mathscr{K}_m(\pi_n z_0 + f)\| \leq \|\mathscr{K}(z_0 + f) - \mathscr{K}_m(z_0 + f)\| \\ + \|\mathscr{K}_m(z_0 + f) - \mathscr{K}_m(\pi_n z_0 + f)\|.\end{aligned} \tag{51}$$

Let

$$\mathscr{K}_m(z_0 + f) - \mathscr{K}_m(\pi_n z_0 + f) = \mathscr{K}_m'(x_0)(z_0 - \pi_n z_0) + O(\|z_0 - \pi_n z_0\|^2).$$

The last term leads to an error of size $O(h^{2r})$. The first term is obtained from Proposition 1. Now, combining (51) with (43) gives the desired result.

The following asymptotic expansion can be proved

$$\tilde{y}_n = x_0 + \eta_1 h^{2r} + O(\max\{h^{2r+2}, \tilde{h}^d\}), \tag{52}$$

where η_1 is independent of h. We can apply the *Richardson* extrapolation and obtain approximations of x_0 of higher order. Define

$$y_n^R = \frac{2^{2r}\tilde{y}_{2n} - \tilde{y}_n}{2^{2r} - 1}.$$

Then we have the following estimate

$$\|y_n^R - x_0\|_\infty = O(\max\{h^{2r+2}, \tilde{h}^d\}). \tag{53}$$

If $\tilde{h}$ and d are chosen such that $\tilde{h} \leq h^{2r+2}$, then

$$\|y_n^R - x_0\|_\infty = O(h^{2r+2}). \tag{54}$$

5 Numerical Results

In this section, numerical examples are given to illustrate the theory established in the previous sections. Let $\mathbb{X}_n$ be the space of piecewise constant functions ($r = 1$) with respect to the uniform partition of $[0, 1]$

$$0 = \frac{1}{n} < \frac{2}{n} < \ldots < \frac{n}{n} = 1.$$

The projection π_n is chosen to be the interpolatory projection at the $nr = n$ midpoints

$$t_i^{(n)} = \frac{2i-1}{2n}, \; i = 1, \ldots, n$$

Let

$$\|x_0 - y_n\|_\infty = O(h^\alpha), \quad \|x_0 - \widetilde{y}_n\|_\infty = O(h^\beta), \quad \|x_0 - y_n^R\|_\infty = O(h^\gamma).$$

Note that, for evaluating the required integrals we use the composite 2 points Gaussian quadrature with respect to the uniform partition of [0, 1] with $m = 128$ intervals. The computations are done for $n = 2, 4, 8, 16, 32$ and 64. Thus,

$$r = 1, \quad d = 4, \quad \tilde{h} = 2^{-7}, \quad h \geq 2^{-6} \;\text{ hence }\; \tilde{h}^d = 2^{-28} \leq 2^{-24} \leq h^{2r+2}.$$

The expected orders of convergence are

$$\alpha = 1, \quad \beta = 2, \quad \gamma = 4.$$

Example 1 We consider the *Hammerstein* equation with a degenerate kernel

$$x(s) - \int_0^1 \cos(\pi s)\sin(\pi t)[x(t)]^2 dt = f(s) \quad s \in [0, 1],$$

where $f \in C[0, 1]$ is selected so that $x_0(s) = |s - \frac{1}{2}|^{\frac{3}{2}}$. The results are given in the Table 1.

The above table illustrate that a high accuracy is obtained by the extrapolated *Kantorovich* method even when the solution and the right hand side are only continuous.

Example 2 Consider

$$x(s) - \int_0^1 \frac{1}{s + t + x(t)} dt = f(s) \quad s \in [0, 1], \tag{55}$$

Table 1 *Kantorovich*-collocation method

n	$\|x_0 - y_n\|_\infty$	α	$\|x_0 - \widetilde{y}_n\|_\infty$	β	$\|x_0 - y_n^R\|_\infty$	γ
2	6.55×10^{-3}	–	6.52×10^{-6}	–	2.65×10^{-7}	–
4	3.54×10^{-3}	0.89	1.83×10^{-6}	1.93	1.07×10^{-8}	4.63
8	1.81×10^{-3}	0.97	4.65×10^{-7}	1.97	5.83×10^{-10}	4.20
16	9.08×10^{-4}	0.99	1.17×10^{-7}	1.99	3.51×10^{-11}	4.05
32	4.54×10^{-4}	1.00	2.92×10^{-8}	2.00		
64	2.27×10^{-4}	1.00				

Table 2 *Kantorovich*-collocation method

n	$\|x_0 - y_n\|_\infty$	α	$\|x_0 - \widetilde{y}_n\|_\infty$	β	$\|x_0 - y_n^R\|_\infty$	γ
2	1.46×10^{-1}	–	8.84×10^{-3}	–	2.36×10^{-5}	–
4	8.13×10^{-2}	0.84	2.19×10^{-3}	2.01	2.06×10^{-6}	3.52
8	4.32×10^{-2}	0.91	5.46×10^{-4}	2.00	1.36×10^{-7}	3.92
16	2.22×10^{-2}	0.95	1.37×10^{-4}	2.00	8.64×10^{-9}	3.98
32	1.13×10^{-2}	0.98	3.41×10^{-5}	2.00		
64	5.71×10^{-3}	0.99				

where f is so chosen that $x_0(s) = \frac{1}{s+1}$ is a solution of (55). The results are given in Table 2.

It can be seen from the above tables that the computed orders of convergence math well with the theoretical ones.

References

1. Allouch, C., Sbibih, D., Tahrichi, M.: Richardson extrapolation of superconvergent projection-type methods for hammerstein equations. Numer. Func. Anal. Optim. **41**(7) (2020)
2. Allouch, C., Sbibih, D., Tahrichi, M.: Superconvergent Nyström method for Urysohn integral equations. BIT Numer. Math. **57**, 3–20 (2017)
3. Atkinson, K.E., Flores, J.: The discrete collocation method for nonlinear integral equations. IMA J. Numer. Anal. **13**, 195–213 (1993)
4. Atkinson, K., Potra, F.: Projection and iterated projection methods for nonlinear integral equations. SIAM J. Numer. Anal. **24**, 1352–1373 (1987)
5. de Boor, C., Swartz, B.: Collocation at Gaussian points. SIAM J. Numer. Anal. **10**, 582–606 (1973)
6. Grammont, L., Kulkarni, R.P., Vasconcelos, P.B.: Modified projection and the iterated modified projection methods for nonlinear integral equations. J. Int. Eq. Appl. **25**(4), 481–516 (2013)
7. Guoqiang, H.: Extrapolation of a discrete collocation-type method of Hammerstein equations. J. Comput. Appl. Math. **61**, 73–86 (1995)
8. Kulkarni, R.P., Nidhin, T.J.: Asymptotic error analysis of projection and modified projection methods for nonlinear integral equations. J. Integr. Eq. Appl. **27**(1), 67–101 (2015)
9. Kulkarni, R.P., Rakshit, G.: Richardson extrapolation for the discrete iterated modified projection solution. Numer. Algor. https://doi.org/10.1007/s11075-019-00808-5
10. Kulkarni, R.P., Rakshit, G.: Discrete modified projection method for Urysohn integral equations with smooth kernels. Appl. Numer. Math. **126**, 180–198 (2018)
11. Kulkarni, R.P., Nidhin, T.J.: Asymptotic error analysis of projection and modified projection methods for nonlinear integral equations. J. Int. Eq. Appl. **27**, 67–101 (2015)
12. Lin, Q., Sloan, I.H., Xie, R.: Extrapolation of the iterated-collocation method for integral equations of the second kind. SIAM J. Numer. Anal. **6**, 1535–1541 (1990)
13. Mclean, W.: Aymptotic error expansions for numerical solution of integral equations. IMA J. Numer. Anal. **9**, 373–384 (1989)
14. Schock, E.: Galerkin-like methods for equations of the second kind. J. Int. Eq. Appl **4**, 361–364 (1982)

15. Sloan, I.H.: Error analysis for a class of degenerate kernel methods. Numer. Math. **25**, 231–238 (1976)
16. Vainikko, G.M.: Galerkin's perturbation method and the general theory of approximate methods for non-linear equations. USSR Comput. Math. Math. Phys. **7**, 1–41 (1967)

The Maximal Numerical Range of a Quadratic Matrix

El Hassan Benabdi

Abstract Let n be a positive integer and let $M_n(\mathbb{C})$ denote the algebra of all complex n-by-n matrices. A matrix $A \in M_n(\mathbb{C})$ is called quadratic if it satisfies some non-trivial quadratic equation $(A - \alpha I)(A - \beta I) = 0$, where I denotes the $n \times n$ identity matrix. In this paper, we give an explicit formula for the maximal numerical range of quadratic matrices.

1 Introduction

Before stating the results, we recall some results from the literature.
Let n be a positive integer and let $\mathbb{C}^n$ stand for the standard n-dimensional inner product space over the complex field $\mathbb{C}$. Denote by $M_n(\mathbb{C})$ the algebra of all complex n-by-n matrices. For $A \in M_n(\mathbb{C})$, the numerical range of A is defined as the set

$$W(A) = \{x^*Ax : x \in \mathbb{C}^n,\ \|x\| = 1\}.$$

It is a celebrated result due to Toeplitz-Hausdorff that $W(A)$ is a convex set in the complex plane. The numerical range of a matrix in $M_n(\mathbb{C})$ is closed. For more details about the theory of numerical range, the reader is referred to [3, 4] and references therein. There is another set that is close to the numerical range $W(A)$; that is the maximal numerical range $W_0(A)$ of A. It was introduced by Stampfli [8] and defined as follows.

Definition 1 For $A \in M_n(\mathbb{C})$, the maximal numerical range $W_0(A)$ of A is given by

$$W_0(A) = \{x^*Ax : x \in \mathbb{C}^n,\ \|x\| = 1,\ \|Ax\| = \|A\|\}.$$

E. H. Benabdi (✉)
Department of Mathematics, Laboratory of Mathematics, Statistics and Applications, Faculty of Sciences, Mohammed V University in Rabat, Rabat, Morocco
e-mail: e.benabdi@um5r.ac.ma

S. Melliani et al. (eds.), *Applied Mathematics and Modelling in Finance, Marketing and Economics*, Studies in Computational Intelligence 1114,
https://doi.org/10.1007/978-3-031-42847-0_6

It was shown in [8] that $W_0(A)$ is nonempty, closed, convex and contained in the numerical range; $W_0(A) \subseteq W(A)$. Note that the notion of the maximal numerical range was introduced by Stampfli [8] (especially) for the purpose of calculating the norm of the inner derivations. Recall that the inner derivation δ_A associated with $A \in M_n(\mathbb{C})$ is defined by

$$\delta_A : M_n(\mathbb{C}) \longrightarrow M_n(\mathbb{C}), \; X \longmapsto AX - XA.$$

Indeed, the author [8] established the following.
For any $A \in M_n(\mathbb{C})$

$$\|\delta_A\| = 2\,\|A - c_A I\|,$$

where c_A is the unique scalar c_A satisfying

$$\|A - c_A I\| = \inf_{\lambda \in \mathbb{C}} \|A - \lambda I\|.$$

The scalar c_A is called the center of mass of A.

Recall that a matrix $A \in M_n(\mathbb{C})$ is called quadratic if it satisfies some non-trivial quadratic equation $(A - \alpha I)(A - \beta I) = 0$, where $\alpha, \beta \in \mathbb{C}$. We denote by $Re(\lambda)$ the real part of $\lambda \in \mathbb{C}$. We have the following.

Theorem 1 ([1, 9]) *Let $A \in M_n(\mathbb{C})$ be a quadratic matrix satisfying $(A - \alpha I)(A - \beta I) = 0$ for some scalars α and β. Then*

(a) *A is unitarily equivalent to a matrix of the form*

$$\alpha I_1 \oplus \beta I_2 \oplus \begin{bmatrix} \alpha I_3 & T \\ 0 & \beta I_3 \end{bmatrix} \text{ on } \mathscr{H}_1 \oplus \mathscr{H}_2 \oplus (\mathscr{H}_3 \oplus \mathscr{H}_3),$$

where $\mathscr{H}_1$, $\mathscr{H}_2$ and $\mathscr{H}_3$ are complex subspaces of $\mathbb{C}^n$ with T is positive definite on $\mathscr{H}_3$.

(b)

$$\begin{aligned} \|A\| = \left\| \begin{bmatrix} \alpha I_3 & T \\ 0 & \beta I_3 \end{bmatrix} \right\| &= \left\| \begin{bmatrix} \alpha & \|T\| \\ 0 & \beta \end{bmatrix} \right\| \\ &= \frac{1}{\sqrt{2}} \sqrt{u + \sqrt{u^2 - v}}, \end{aligned}$$

where $u = |\alpha|^2 + |\beta|^2 + \|T\|^2$ and $v = 4\,|\alpha|^2\,|\beta|^2$.

Proposition 1 ([1]) *Let $A \in M_n(\mathbb{C})$ be a quadratic matrix satisfying $(A - \alpha I)(A - \beta I) = 0$ for some scalars α and β. Then, the center of mass of A is*

$$c_A = \frac{\alpha + \beta}{2}.$$

Theorem 2 ([5]) *Let* $A = \begin{bmatrix} \alpha & \gamma \\ 0 & \beta \end{bmatrix}$, *where* $\alpha, \beta, \gamma \in \mathbb{C}$. *Then*

$$\begin{cases} W_0(A) = \left\{ \dfrac{\|A\|^2(\alpha+\beta) - \alpha\beta(\overline{\alpha}+\overline{\beta})}{2\|A\|^2 - |\alpha|^2 - |\beta|^2 - |\gamma|^2} \right\}, & \text{if } \gamma \neq 0 \text{ or } |\alpha| \neq |\beta|; \\ W_0(A) = [\alpha, \beta], & otherwise. \end{cases}$$

In Sect. 2, we provide an explicit formula for the maximal numerical range of a quadratic matrix using the fact that a quadratic matrix is unitarily equivalent to a direct sum of matrices relatively well-known.

2 Maximal Numerical Range of a Quadratic Matrix

In this section, we calculate the maximal numerical rang of a quadratic matrix. Let $A \in M_n(\mathbb{C})$ be a quadratic matrix satisfying the following quadratic equation $(A - \alpha I)(A - \beta I) = 0$, where $\alpha, \beta \in \mathbb{C}$. From Theorem 1, there exist subspaces $\mathcal{H}_1$, $\mathcal{H}_2$ and $\mathcal{H}_3$ of $\mathbb{C}^n$ such that A is unitarily equivalent to a matrix of the form

$$\alpha I_1 \oplus \beta I_2 \oplus \begin{bmatrix} \alpha I_3 & T \\ 0 & \beta I_3 \end{bmatrix} \text{ on } \mathcal{H}_1 \oplus \mathcal{H}_2 \oplus (\mathcal{H}_3 \oplus \mathcal{H}_3),$$

where T is positive definite on $\mathcal{H}_3$. According to [6, Lemma 2],

$$W_0(A) = W_0\left(\begin{bmatrix} \alpha I_3 & T \\ 0 & \beta I_3 \end{bmatrix}\right).$$

Theorem 3 *Let* $A \in M_n(\mathbb{C})$ *be a quadratic matrix satisfying* $(A - \alpha I)(A - \beta I) = 0$ *for some scalars* α *and* β *and let* T *be the positive definite matrix such that* A *is unitarily equivalent to* $\alpha I_1 \oplus \beta I_2 \oplus \begin{bmatrix} \alpha I_3 & T \\ 0 & \beta I_3 \end{bmatrix}$. *Then*

$$\begin{cases} W_0(A) = \left\{ \dfrac{\|A\|^2(\alpha+\beta) - \alpha\beta(\overline{\alpha}+\overline{\beta})}{2\|A\|^2 - |\alpha|^2 - |\beta|^2 - \|T\|^2} \right\}, & \text{if } T \neq 0 \text{ or } |\alpha| \neq |\beta|; \\ W_0(A) = [\alpha, \beta], & otherwise. \end{cases}$$

Proof We show that $W_0(A) = W_0\left(\begin{bmatrix} \alpha & \|T\| \\ 0 & \beta \end{bmatrix}\right)$ and we then conclude by Theorem 2. If $T = 0$, the result is clear. If $T \neq 0$ and $\alpha = 0$, by Theorem 2,

$$W_0\left(\begin{bmatrix}\alpha & \|T\| \\ 0 & \beta\end{bmatrix}\right) = \{\beta\}.$$

We also have $W_0(A) = \{\beta\}$. Indeed, let $\lambda \in W_0(A)$, then there is $x = y \oplus z \in \mathscr{H}_3 \oplus \mathscr{H}_3$ with $\|y\|^2 + \|z\|^2 = 1$ such that $x^* Ax = \lambda$ and $\|Ax\|^2 = \|A\|^2 = \|T\|^2 + |\beta|^2$. Since $\|Ax\|^2 = \|Tz\|^2 + |\beta|^2\|z\|^2$, then $\|z\| = 1$ and $\|y\| = 0$. We derive that $x^*Ax = \big(y^*Tz + \beta|z|^2\big) = \beta$. Therefore, we may assume that $T \neq 0$ and $\alpha \neq 0$.

We claim that $W_0(A) \subseteq W_0\left(\begin{bmatrix}\alpha & \|T\| \\ 0 & \beta\end{bmatrix}\right)$.

Let $\lambda \in W_0(A)$, then there exists a unit vector x in $\mathscr{H}_3 \oplus \mathscr{H}_3$ such that $\|Ax\| = \|A\|$ and $x^*Ax = \lambda$. We decompose x as $ay \oplus bz$ where $|a|^2 + |b|^2 = 1$ and $\|y\| = \|z\| = 1$. Note that we can assume that $\alpha a\overline{b} \geq 0$. Indeed, let θ_α, θ_a and θ_b be the arguments of α, a and b, respectively. Set $a' = ae^{-i\theta_a}$, $b' = be^{-i(\theta_b - \theta_\alpha)}$, $y' = e^{i\theta_a}y$ and $z' = e^{i(\theta_b - \theta_\alpha)}z$. It is clear that $|a'|^2 + |b'|^2 = 1$, $\|y'\| = \|z'\| = 1$, $x = a'y' \oplus b'z'$ and it is easy to see that $\alpha a'\overline{b'} = |\alpha ab| \geq 0$. Therefore, we have

$$\begin{aligned}
\|Ax\|^2 &= |\alpha|^2|a|^2 + 2\alpha a\overline{b}Re\big(y^*Tz\big) + |b|^2\|Tz\|^2 + |\beta|^2|b|^2 \\
&\leq |\alpha|^2|a|^2 + 2\alpha a\overline{b}\|T\| + |b|^2\|T\|^2 + |\beta|^2|b|^2 \\
&= \left\|\begin{bmatrix}\alpha & \|T\| \\ 0 & \beta\end{bmatrix}\begin{bmatrix}a \\ b\end{bmatrix}\right\|^2 \\
&\leq \left\|\begin{bmatrix}\alpha & \|T\| \\ 0 & \beta\end{bmatrix}\right\|^2 \\
&= \|A\|^2 \quad \big(\text{by Theorem 1(b)}\big).
\end{aligned}$$

Since $\|Ax\| = \|A\|$, we derive that

$$\left\|\begin{bmatrix}\alpha & \|T\| \\ 0 & \beta\end{bmatrix}\begin{bmatrix}a \\ b\end{bmatrix}\right\| = \left\|\begin{bmatrix}\alpha & \|T\| \\ 0 & \beta\end{bmatrix}\right\|.$$

A simple computation shows that

$$x^*Ax = \alpha|a|^2 + b\overline{a}(y^*Tz) + \beta|b|^2$$

and

$$\begin{bmatrix}a \\ b\end{bmatrix}^*\begin{bmatrix}\alpha & \|T\| \\ 0 & \beta\end{bmatrix}\begin{bmatrix}a \\ b\end{bmatrix} = \alpha|a|^2 + b\overline{a}\|T\| + \beta|b|^2.$$

If $b\overline{a} = 0$, then $x^*Ax = \begin{bmatrix}a \\ b\end{bmatrix}^*\begin{bmatrix}\alpha & \|T\| \\ 0 & \beta\end{bmatrix}\begin{bmatrix}a \\ b\end{bmatrix}$, so $\lambda \in W_0\left(\begin{bmatrix}\alpha & \|T\| \\ 0 & \beta\end{bmatrix}\right)$. If $b\overline{a} \neq 0$, since $a\overline{b}\Big(Re\big(y^*Tz\big) - \|T\|\Big) = 0$, then $Re\big(y^*Tz\big) = \|T\|$. This implies $y^*Tz =$

$\|T\|$ and, as above, we again have $\lambda \in W_0\left(\begin{bmatrix}\alpha & \|T\| \\ 0 & \beta\end{bmatrix}\right)$. Consequently, $W_0(A) \subseteq W_0\left(\begin{bmatrix}\alpha & \|T\| \\ 0 & \beta\end{bmatrix}\right)$.

We now claim that $W_0\left(\begin{bmatrix}\alpha & \|T\| \\ 0 & \beta\end{bmatrix}\right) \subseteq W_0(A)$.

Let $\lambda \in W_0\left(\begin{bmatrix}\alpha & \|T\| \\ 0 & \beta\end{bmatrix}\right)$, then there exist $a, b \in \mathbb{C}$ such that $|a|^2 + |b|^2 = 1$,

$$\left\|\begin{bmatrix}\alpha & \|T\| \\ 0 & \beta\end{bmatrix}\begin{bmatrix}a \\ b\end{bmatrix}\right\| = \left\|\begin{bmatrix}\alpha & \|T\| \\ 0 & \beta\end{bmatrix}\right\| \text{ and } \begin{bmatrix}a \\ b\end{bmatrix}^* \begin{bmatrix}\alpha & \|T\| \\ 0 & \beta\end{bmatrix}\begin{bmatrix}a \\ b\end{bmatrix} = \lambda.$$

Let z be a unit vector in $\mathscr{H}_3$ such that $\|Tz\| = \|T\|$. Set $y := Tz/\|Tz\|$ and $x := ay \oplus bz$. We have

$$\|Ax\|^2 = |\alpha|^2|a|^2 + 2Re\big(\alpha a\overline{b}\big)\|Tz\| + |b|^2\|Tz\|^2 + |\beta|^2|b|^2$$

and

$$\left\|\begin{bmatrix}\alpha & \|T\| \\ 0 & \beta\end{bmatrix}\begin{bmatrix}a \\ b\end{bmatrix}\right\|^2 = |\alpha|^2|a|^2 + 2Re\big(\alpha a\overline{b}\big)\|T\| + |b|^2\|T\|^2 + |\beta|^2|b|^2.$$

Hence

$$\|Ax\| = \left\|\begin{bmatrix}\alpha & \|T\| \\ 0 & \beta\end{bmatrix}\begin{bmatrix}a \\ b\end{bmatrix}\right\| = \left\|\begin{bmatrix}\alpha & \|T\| \\ 0 & \beta\end{bmatrix}\right\| = \|A\|.$$

On the other hand,

$$x^*Ax = \alpha|a|^2 + b\overline{a}\,\|Tz\| + \beta|b|^2$$

and

$$\begin{bmatrix}a \\ b\end{bmatrix}^* \begin{bmatrix}\alpha & \|T\| \\ 0 & \beta\end{bmatrix}\begin{bmatrix}a \\ b\end{bmatrix} = \alpha|a|^2 + b\overline{a}\|T\| + \beta|b|^2.$$

We derive that

$$x^*Ax = \begin{bmatrix}a \\ b\end{bmatrix}^* \begin{bmatrix}\alpha & \|T\| \\ 0 & \beta\end{bmatrix}\begin{bmatrix}a \\ b\end{bmatrix} = \lambda.$$

It follows that $\lambda \in W_0(A)$. Thus, $W_0\left(\begin{bmatrix}\alpha & \|T\| \\ 0 & \beta\end{bmatrix}\right) \subseteq W_0(A)$. In summary, $W_0(A) = W_0\left(\begin{bmatrix}\alpha & \|T\| \\ 0 & \beta\end{bmatrix}\right)$. This completes the proof.

For a bounded linear operator T on a complex Banach space, let $\sigma(T)$ denote the spectrum of T and $\sigma_n(T) := \{\lambda \in \sigma(T) : |\lambda| = \|T\|\}$.

Remark 1 The result of [6, Lemma 2] does not hold in the infinite case. Indeed, let A_k for $k = 1, 2, \ldots$, be the 2-by-2 matrix:

$$A_k = \begin{bmatrix} 2 & 0 \\ 0 & \frac{1}{k} - 2 \end{bmatrix}.$$

It is known that $\sigma\big(\oplus_k A_k\big) = \overline{\cup_k \sigma(A_k)}$. That is, $\overline{\cup_k\{\frac{1}{k} - 2, 2\}} = \sigma\big(\oplus_k A_k\big)$. Since $\|\oplus_k A_k\| = 2$, $\sigma_n\big(\oplus_k A_k\big) = \{-2, 2\}$. We derive that $W_0(\oplus_k A_k) = [-2, 2]$; see [2, 7]. But, $W_0(A_k) = \{2\}$, for $k = 1, 2, \ldots$, then $\cup_k W_0(A_k) = \{2\}$. Consequently, $W_0(\oplus_k A_k) \neq \cup_k W_0(A_k)$.

References

1. Abu-Omar, A., Wu, P.Y.: Scalar approximants of quadratic operators with applications. Oper. Matrices **12**(1), 253–262 (2018)
2. Baghdad, A., Kaadoud, M.C.: On the maximal numerical range of a hyponormal operator. Oper. Matrices **13**(4), 1163–1171 (2019)
3. Gustafson, K.E., Rao, D.K.M.: Numerical Range: The Field of Values of Linear Operators and Matrices. Springer, New York (1997)
4. Halmos, P.R.: Hilbert Space Problem Book. Van Nostrand, New York (1967)
5. Hamed, A.N., Spitkovsky, I.M.: On the maximal numerical range of some matrices. Electron. J. Linear Algebra **34**, 288–303 (2018)
6. Ji, G., Liu, N., Li, Z.: Essential numerical range and maximal numerical range of the Aluthge transform. Linear Multilinear Algebra **55**(4), 315–322 (2007)
7. Spitkovsky, I.M.: A note on the maximal numerical range. Oper. Matrices **13**(3), 601–605 (2019)
8. Stampfli, J.G.: The norm of derivation. Pacific J. Math. **33**, 737–747 (1970)
9. Tso, S.H., Wu, P.Y.: Matricial ranges of quadratic operators. Rocky Mountain J. Math. **29**, 1139–1152 (1999)

The Effect of Change in Basilar Membrane Stiffness on the Micromechanics Cochlear Model

F. Kouilily, F. E. Aboulkhouatem, N. Yousfi, N. Achtaich, and M. El Khasmi

Abstract In this present work, the micromechanical cochlea model has developed in order to describe mathematically the displacement of cochlear partition using finite difference method and Cramer's rule, Then, we have studied the effect of basilar membrane (BM) stiffness on the displacements of the BM and tectorial membrane (TM). Results showed that the augmentation of the BM stiffness reduce the maximum amplitude displacement of the BM and TM. These findings contribute to understand that the loss of hearing at low frequencies may be the result of altered cochlear micromechanics.

1 Introduction

The cochlea is the organ of hearing system, where acoustic signals are converted into nerve impulses, before conveyed to the brain. The first cochlear model consists of two uncoiled compartments filled with fluid which are divided by the BM [1–3]. In such model, by considering the pressure difference across the BM, this membrane is represented by one degree of freedom system. Experiments revealed that the TM has many mechanical proprieties which can be important in the cochlear response and in its propagation waves [4, 5]. The motions of BM and TM are coupled by the damping c_3 and stiffness k_3 that represented by the Organ of Corti (OC) which included the outer hair cell (OHC) (see Fig. 1) [6]. To demonstrate the role of TM in the response of cochlea, the models of one degree of freedom are generalized into models of two degrees of freedom of cochlear partition (model of Neely and Kim) [7–9]. Analysis of the two degree of freedom model for the human cochlea was given by Ku [10, 11] a satisfactory results respecting the results discovered by Békésy [12]. Therefore, the study of cochlear micromechanics is used and developed to resolve many problems of hearing for otoacoustics emissions [13, 14]. Cochlear

F. Kouilily (✉) · F. E. Aboulkhouatem · N. Yousfi · N. Achtaich · M. El Khasmi
Faculty of Sciences Ben M'sik, Hassan II University, Avenue Cdt, Driss El Harti B.P: 7955, Casablanca, Morocco
e-mail: kouililyfatiha@gmail.com

S. Melliani et al. (eds.), *Applied Mathematics and Modelling in Finance, Marketing and Economics*, Studies in Computational Intelligence 1114,
https://doi.org/10.1007/978-3-031-42847-0_7

hearing is the most common type of hearing loss. Its produced when the organs inside the cochlea are damaged due to noise, age, or in its pathological structure. One of cases of pathological structure is the abnormality in the function of the BM resulting from various illness that affect the response of the cochlea [15–17], such as Alport syndrome [18–20] and Meniere's disease [21, 22]. Therefore, the aim of this study is to propose the mathematical solution of the coupled ordinary differential equations of Neely and Kim by using finite difference method and Cramer's rule, then we study the changes observed by augmenting the stiffness on the vibratory behavior of the BM and TM. The paper will be organized as follows: firstly, we introduce the description of the cochlear micromechanics model. Secondly, we give the solution of the model. Then, we present the numerical results, showing the effect of the BM stiffness of the coupled response cochlea, and finally, a general conclusion.

2 Model of Cochlear Mechanics

The cochlea is represented as two compartments filled with fluid and separated by the cochlear partition (CP), which contain the BM, TM and the OC as shown in Fig. 1.

The one-dimensional wave propagation equation along the cochlea [23] is given by:

$$\frac{\partial^2 P_d(x,t)}{\partial x^2} = \frac{2\rho}{H}\frac{\partial^2 \xi_p(x,t)}{\partial t^2} \tag{1}$$

where P_d is the pressure difference across the CP. ξ_p is the displacement of CP, ρ and H are the density of the CP and the height of fluid compartment, respectively. At the basal and apical ends of the cochlea, the boundary conditions are given by

$$\frac{\partial P_d(x,t)}{\partial x}|_{x=0} = 2\rho\frac{\partial^2 \xi_s(x,t)}{\partial t^2} \tag{2}$$

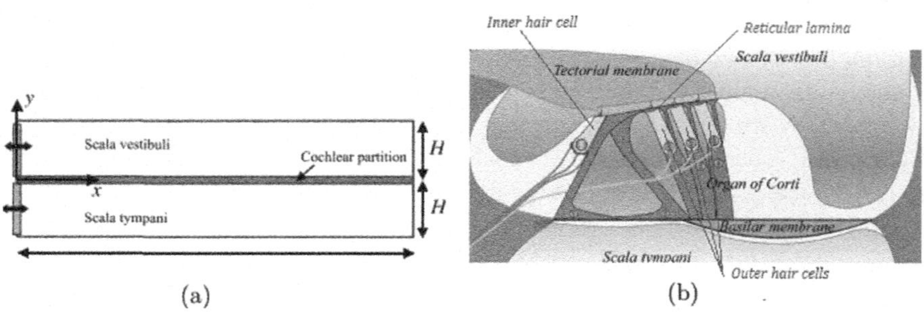

Fig. 1 Structure of cochlea (**a**) Chambers of the cochlea separated by the CP, and (**b**) Zoom of the CP which included the OC, supported by BM and recovered by TM [6]

and

$$P_d(x,t)|_{x=L} = 0 \tag{3}$$

where ξ_s represents the displacement of the stapes.

Each of the displacement variables ξ_p, and ξ_s can be eliminated from the above equations by using a frequency domain and by introducing the appropriate mechanical impedance functions.

By introducing the impedance of the middle ear Z_m

$$Z_m = \frac{k_m}{iw} + c_m + iwm_m$$

Then, the stapes acceleration can be expressed as

$$\ddot{\xi}_s(x) = (\frac{iw}{Z_m})((\frac{A_m}{G_m A_s})P_e - P_d(0)) \tag{4}$$

where A_m and A_s are the effective area respectively, of the eardrum and stapes. G_m is the gain of the middle ear. The CP acceleration can be expressed as

$$\ddot{\xi}_p(x) = (\frac{iw}{Z_p(x)})P_d(x) \tag{5}$$

where $Z_p(x)$

$$Z_p = (\frac{g}{b})(Z_1 + \frac{Z_2(Z_3 - \gamma Z_4)}{(Z_2 + Z_3)})$$

is the partition impedance which describe the coupling between the macromechanics proprieties of the cochlear fluid and the micromechanics structure of the OC [8]. By substituting Eqs. (4) and (5) into Eqs. (1)–(3), we obtain a one dimensional model expressed only by terms of pressure difference $P_d(x)$.

The micromechanics model of Neely and Kim [8] consists of two degree of freedom system, as illustrated in Fig. 2. The two state variables are associated with each degree of freedom which represent the displacement ξ_b and ξ_t of BM and TM, respectively.

The variables m_1, k_1 and c_1 represent the mass, stiffness and damping of BM, respectively. The mass, stiffness and damping of TM are represented by m_2, k_2 and c_2, respectively. The two masses $m_1$and m_2 are coupled by k_3 and c_3 the motion between BM and TM, ξ_c represents displacement of OC, ξ_t is the TM displacement and $\xi_b$indicates the BM displacement. P_a is the active pressure produced by OHC.

The coupled equation of the motion of BM and TM can be written as

$$P_d - P_a = [m_1\ddot{\xi}_b + c_1\dot{\xi}_b + k_1\xi_b] + [c_3\dot{\xi}_c + k_3\xi_c] \tag{6}$$

and

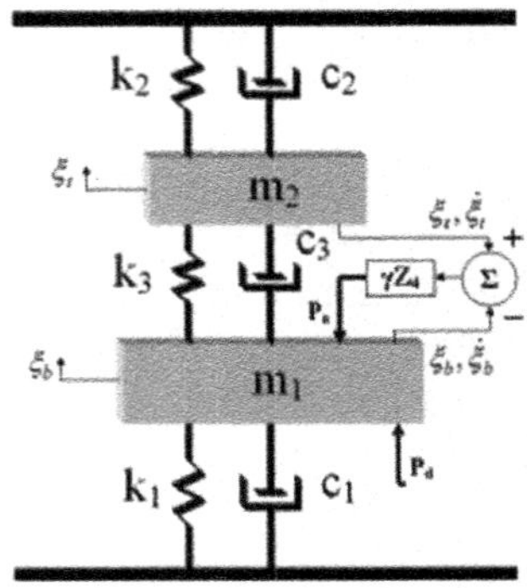

Fig. 2 The Neely and Kim model of the cochlea [8]

$$0 = [m_2\ddot{\xi}_t + c_2\dot{\xi}_t + k_2\xi_t] - [c_3\dot{\xi}_c + k_3\xi_c] \tag{7}$$

where $P_a = -\gamma Z_4(x)\dot{\xi}_c$ is the pressure produced by the OHC, γ is the active gain produced by OHC force generation, $Z_4(x)$ is the impedance associated with the active pressure. where $Z_4 = c_4 + k_4/iw$, and P_a can be expressed as:

$$P_a = -\gamma(c_4\dot{\xi}_c + k_4\xi_c) \tag{8}$$

ξ_c is the difference in position between the BM and the TM,

$$\xi_c = \xi_b - \xi_t \tag{9}$$

Then, from the Eqs. (8) and (9), we have:

$$m_1\ddot{\xi}_b = P_d + (\gamma[c_4(\dot{\xi}_t - \dot{\xi}_b) + k_4(\xi_t - \xi_b)] - \dot{\xi}_b(c_1 + c_3) - \xi_b(k_1 + k_3) + \dot{\xi}_t c_3 + \xi_t k_3) \tag{10}$$

and

$$m_2\ddot{\xi}_t = -\dot{\xi}_t(c_2 + c_3) - \xi_t(k_2 + k_3) + \dot{\xi}_b c_3 + \xi_b k_3 \tag{11}$$

3 Materials and Methods

3.1 *Implementation of Cochlear Mechanics Model Using Finite Difference Method*

Replacing the equation $\frac{\partial^2 P_d}{\partial x^2}$ by its finite difference approximation, Eq. (1) can be written as, from j = 2 to N–1:

$$\frac{P_d(j+1) - 2P_d(j) + P_d(j-1)}{\Delta x^2} - \frac{2iw\rho}{HZ_p(j)}P_d(j) = 0 \tag{12}$$

The boundary condition at the base Eq. (2) can be written using finite difference approximation:

$$\frac{P_d(2) - P_d(1)}{\Delta x} = 2\rho(\frac{iw}{Z_m})((\frac{A_m}{G_m A_s})P_e - P_d(1)) \tag{13}$$

At the apex,

$$P_d(N) = 0 \tag{14}$$

Then, we obtain the following system,

$$\frac{1}{\Delta x}(A - \frac{2\rho wi}{H}B)P_d = C$$

where

$$A = \begin{pmatrix} (\beta-1)\Delta x & \Delta x & & & & 0 \\ 1 & -2 & 1 & & & \\ & \ddots & \ddots & \ddots & & \\ & & \ddots & \ddots & \ddots & \\ & & & 1 & -2 & 1 \\ 0 & & & & 0 & \Delta x^2 \end{pmatrix}, B = \begin{pmatrix} 0 & & & 0 \\ & Y_p(2) & \ddots & \\ & & \ddots & \\ & & \ddots & Y_p(N-1) \\ 0 & & & 0 \end{pmatrix} P_d = \begin{pmatrix} P_d(1) \\ P_d(2) \\ \vdots \\ \vdots \\ P_d(N) \end{pmatrix} et \ C = \begin{pmatrix} \frac{2\rho}{Z_m}\frac{iwA_m}{G_m A_s}P_e \\ 0 \\ \vdots \\ \vdots \\ 0 \end{pmatrix}$$

Or $Y_p = \frac{1}{Z_p}$ and $\beta = 2\rho\Delta x(\frac{iw}{Z_m})$

3.2 *Solution of the Coupled Equations Using Cramer's Rule*

The system of Eqs. (10) and (11) at x can be expressed in matrix form as

$$\begin{pmatrix} m_1 & 0 \\ 0 & m_2 \end{pmatrix}\begin{pmatrix} \ddot{\xi}_b \\ \ddot{\xi}_t \end{pmatrix} + \begin{pmatrix} \gamma c_4 + (c_1 + c_3) & -\gamma c_4 - c_3 \\ -c_3 & (c_2 + c_3) \end{pmatrix}\begin{pmatrix} \dot{\xi}_b \\ \dot{\xi}_t \end{pmatrix} + \begin{pmatrix} \gamma k_4 + (k_1 + k_3) & -\gamma k_4 - k_3 \\ -k_3 & (k_2 + k_3) \end{pmatrix}\begin{pmatrix} \xi_b \\ \xi_t \end{pmatrix} = \begin{pmatrix} P_d \\ 0 \end{pmatrix} \tag{15}$$

The coupled ordinary differential Eqs. (10) and (11) become algebraic equations in the frequency domain

$$\left[-w^2\begin{pmatrix} m_1 & 0 \\ 0 & m_2 \end{pmatrix} + iw\begin{pmatrix} \gamma c_4 + (c_1 + c_3) & -\gamma c_4 - c_3 \\ -c_3 & (c_2 + c_3) \end{pmatrix} + \begin{pmatrix} \gamma k_4 + (k_1 + k_3) & -\gamma k_4 - k_3 \\ -k_3 & (k_2 + k_3) \end{pmatrix}\right]\begin{pmatrix} \widetilde{\xi}_b \\ \widetilde{\xi}_t \end{pmatrix} = \begin{pmatrix} \widetilde{P}_d \\ 0 \end{pmatrix} \tag{16}$$

Using the Cramer's rule, the displacement of $\widetilde{\xi}_b$ and $\widetilde{\xi}_t$ can be expressed as

$$\widetilde{\xi}_b = \frac{\Delta\widetilde{\xi}_b}{\Delta} \tag{17}$$

and

$$\widetilde{\xi}_t = \frac{\Delta\widetilde{\xi}_t}{\Delta} \tag{18}$$

where

$$\Delta = (\gamma(k_4 + iwc_4) + iwc_3 + k_3)(k_2 - w^2 m_2 + iwc_2) + (k_1 - w^2 m_1 + iwc_1)(k_2 + k_3 - w^2 m_2 + iw(c_2 + c_3)),$$

$$\Delta\widetilde{\xi}_b = \widetilde{P}_d(k_2 + k_3 + iw(c_2 + c_3) - w^2 m_2)$$

and

$$\Delta\widetilde{\xi}_t = \widetilde{P}_d(k_3 + iwc_3)$$

Then, the displacement of each degree of freedom is obtained by the real part of the Eqs. (17) and (18).

4 Results and Discussion

4.1 Response of Cochlear Partition Model

The active micromechanical model is solved by using Cramer's rule of algebraic coupled equations. For the evaluation of the results, the numerical solution proposed by Neely and Kim [8] is compared by our results, the model has been tested by simulating the cochlear response for different values of frequencies. Table 1 lists the parameter values for simulating the human cochlea [10].

Figures 3 and 4 show the time courses of the TM and BM displacement respectively for different values of frequencies. The horizontal axis shows the position along the BM and TM along the cochlea. The vertical axis shows the displacement of the BM and TM.

As observed in Figs. 3 and 4, the position where the displacement of the BM and TM attained its maximum displacement changed with respect to the stimulus frequency. For example, the Figs. 3(a) and 4(a) show that the displacement of the BM and TM reached its maximum at the apical part of the BM and BM for low frequencies.

Table 1 Revised parameters of the micromechanical model [10]

Parameters	Values	Units (SI)
$k_1(x)$	$4.95 \times 10^9 \times e^{(-320\times(x+0.00375))}$	$N.m^{-3}$
$c_1(x)$	$1 + 19700 \times e^{(-179\times(x+0.00375))}$	$N.s.m^{-3}$
m_1	1.35×10^{-2}	$Kg.m^{-2}$
$k_2(x)$	$3.15 \times 10^7 \times e^{(-352\times(x+0.00375))}$	$N.m^{-3}$
$c_2(x)$	$113 \times \times e^{(-176\times(x+0.00375))}$	$N.s.m^{-3}$
m_2	2.3×10^{-3}	$Kg.m^{-2}$
$k_3(x)$	$4.5 \times 10^7 \times e^{(-320\times(x+0.00375))}$	$N.m^{-3}$
$c_3(x)$	$22.5 \times e^{(-64\times(x+0.00375))}$	$N.s.m^{-3}$
$k_4(x)$	$2.82 \times 10^9 \times e^{(-320\times(x+0.00375))}$	$N.m^{-3}$
$c_4(x)$	$9650 \times e^{(-164\times(x+0.00375))}$	$N.s.m^{-3}$

where x is the position along the cochlea

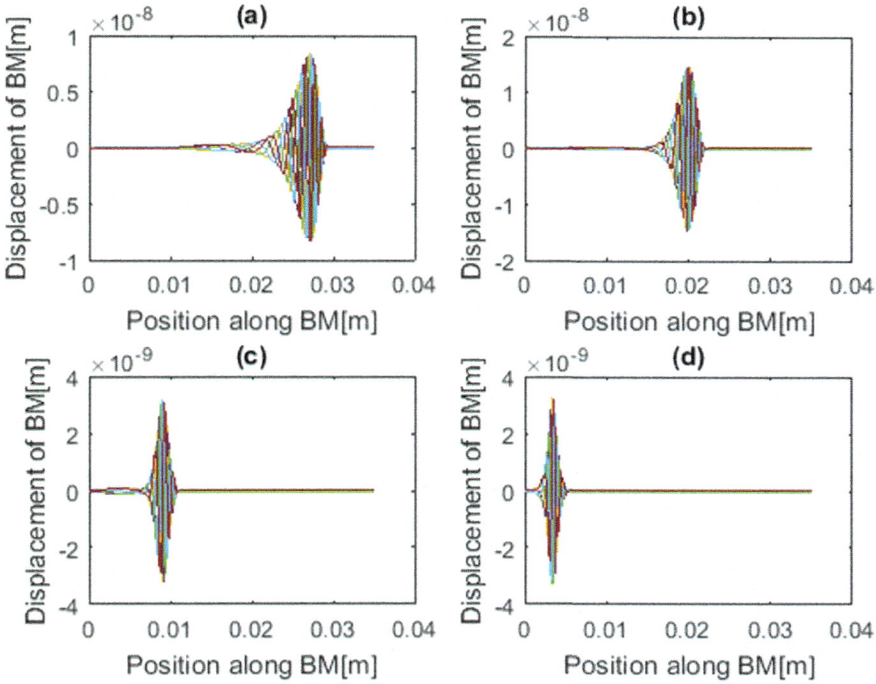

Fig. 3 Time course of the displacement of the BM for various values of frequencies (**a**) f = 440 Hz, (**b**) f = 1220 Hz, (**c**) f = 5620 Hz and (**d**) f = 126880 Hz

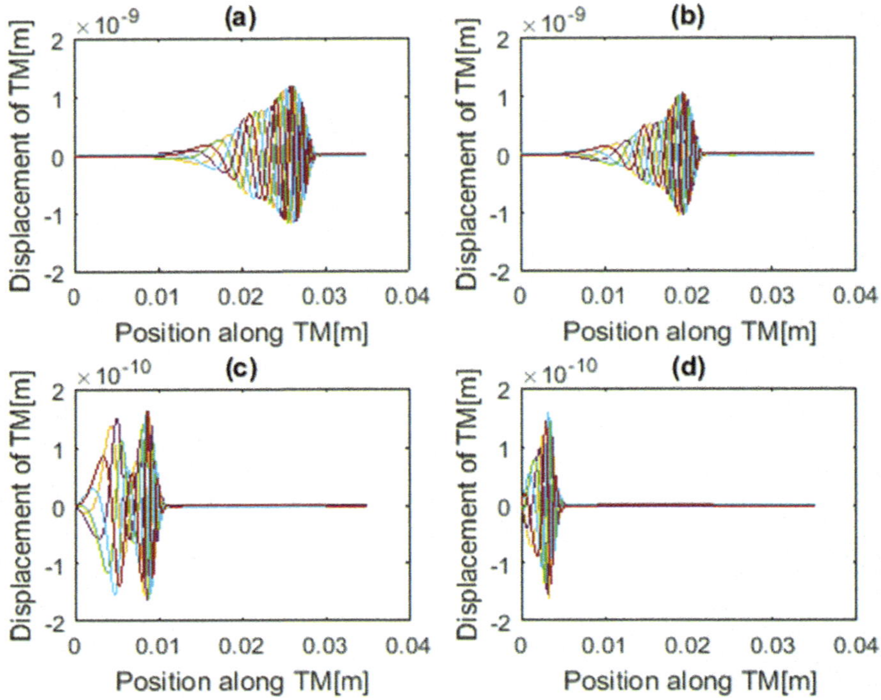

Fig. 4 Time course of the displacement of the TM for various values of frequencies **a** f = 440 Hz, **b** f = 1220 Hz, **c** f = 5620 Hz and **d** f = 12880 Hz

On the other hand, the result of high frequencies shows the maximum displacement at the more basal part of the BM and TM. These results mean that the BM and TM have a frequency discrimination ability.

Figure 5 represents the comparison between our solution using Cramer's rule and Neely and Kim solution for $\gamma = 0$ and $\gamma = 1$. The results of these two different methodologies are similar. When $\gamma = 0$, the model shows the response of a passive cochlea, and when $\gamma = 1$ the model generally shows larger amplification which means the active cochlea. The changing of the active gain γ is used to represent the effect of OHC.

5 The Effect of BM Stiffness on the Displacement of the CP

The increase of stiffness is applied to the parameter k_1 of the BM, so as to show the influence of stiffer BM on the response of the cochlea. The displacement of the BM and TM caused by an increase of BM stiffness (abnormal case) is smaller than that of the case without perturbation (normal case).

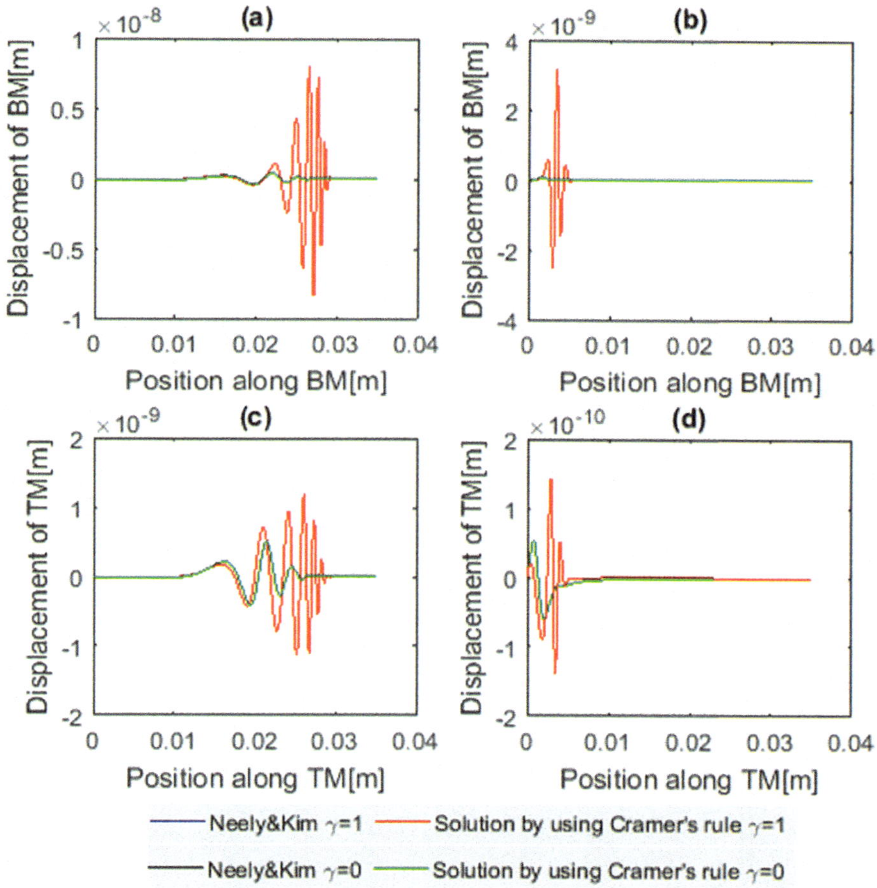

Fig. 5 Comparison between Neely & Kim and Cramer's rule solutions (**a**) The displacement of BM when f = 440 Hz, (**b**) The displacement of BM when f = 12880 Hz, (**c**) The displacement of TM when f = 440 Hz and (**d**) The displacement of TM when f = 12880 Hz

The principal changes observed in comparing the two cases (normal and abnormal displacements of CP) of the response of CP show that the maximum displacement of the BM and the TM is decreased, when the value of BM stiffness is increased.

As shown in Figs. 6 and 8, the reduce in the maximum displacement of the BM and TM between the two cases is more remarkable for low frequencies than high frequencies (Figs. 7 and 9). Figure 10 represents the graph of the changing observed in the maximum displacement of the BM and the TM during applying an increase of BM stiffness ($k_1 + 10^6$) for different values of frequencies. Figure 11 shows the graph of change observed in the position along the CP according to frequency for an increase of BM stiffness.

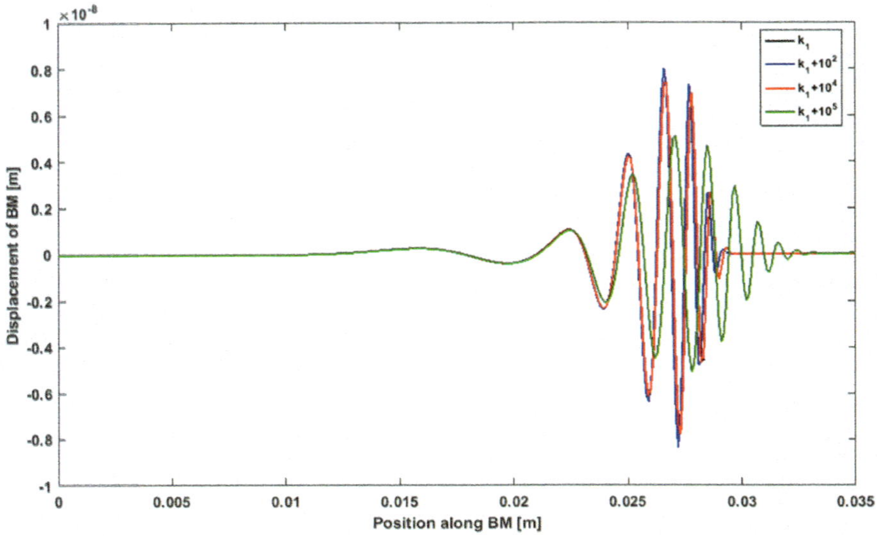

Fig. 6 The effect of BM stiffness on its displacement for f = 440 Hz

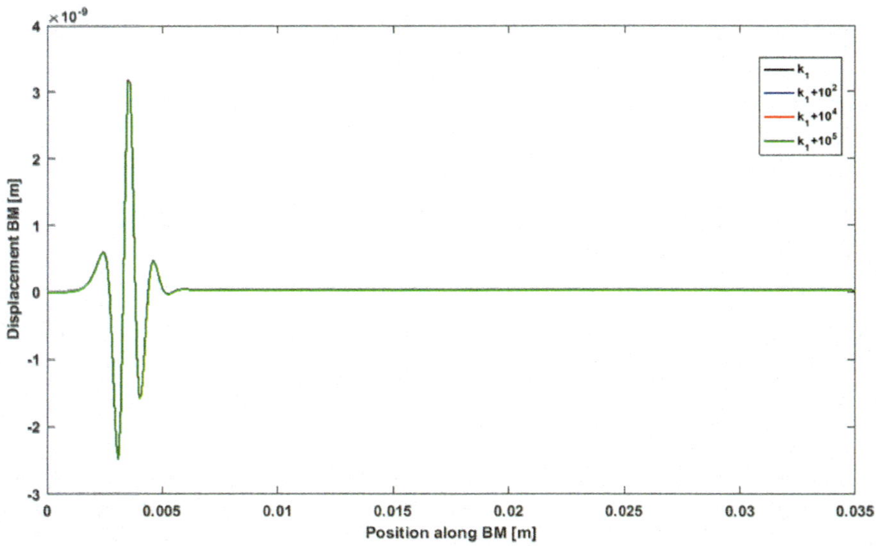

Fig. 7 The effect of BM stiffness on its displacement for f = 12880 Hz

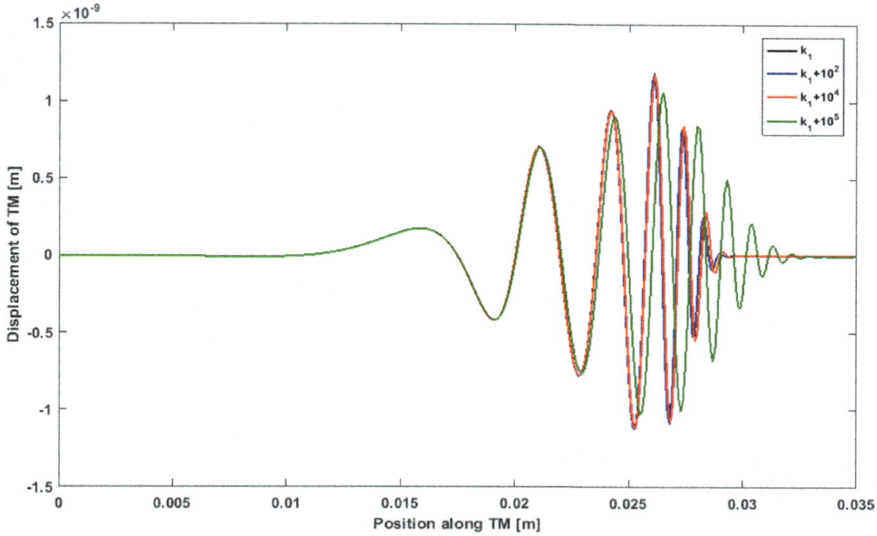

Fig. 8 The effect of BM stiffness on the displacement of the TM for f = 440 Hz

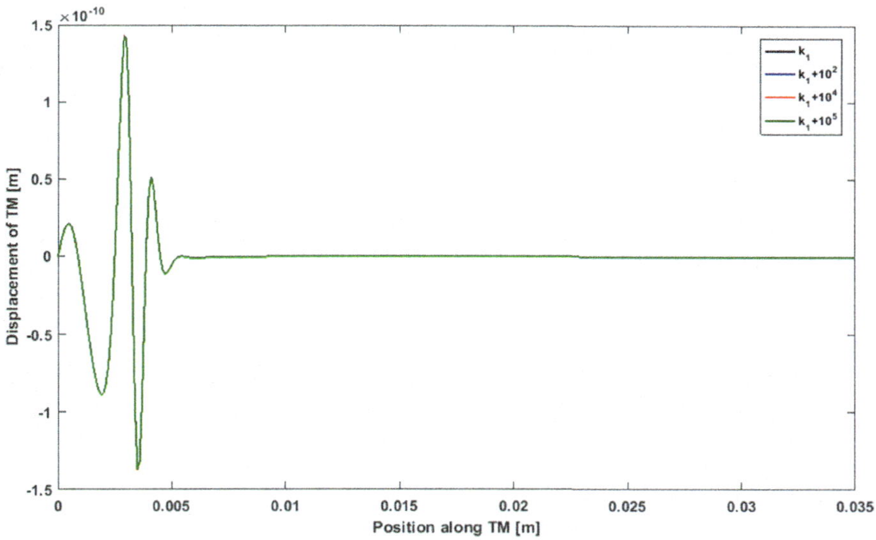

Fig. 9 The effect of BM stiffness on the displacement of the TM for f = 12880 Hz

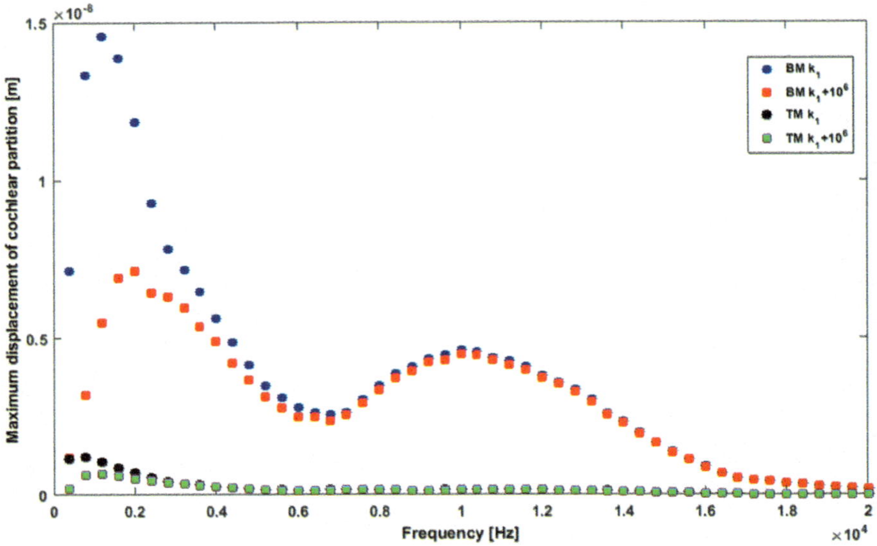

Fig. 10 The Effect of BM stiffness on the response of CP: Peak frequency of the BM and TM versus its maximum displacement

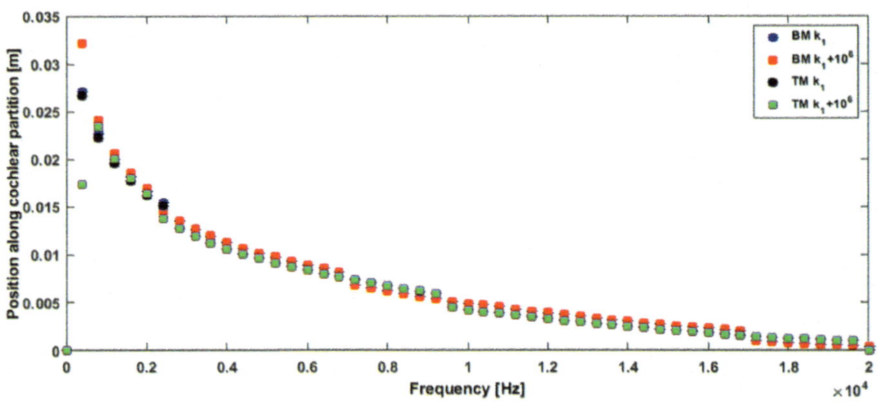

Fig. 11 The Effect of BM stiffness on the response of CP: Peak frequency of the BM and TM versus its position

The maximum displacement of the two membranes is gradually decreased with an increase of BM stiffness, this change in the maximum displacement of the BM and TM is significantly observed especially at low frequencies.
The apex area of the two membranes is highly perturbed, for example, at low frequency the position where the displacement of the BM and TM achieved its maximum displacement is changed. The change of characteristic frequency(CF) location of low

frequencies propose that the frequency discrimination ability can be decreased. This result shows that the degree of an increase of BM stiffness might be could influence hearing ability.

6 Conclusion

The micromechanical model of the cochlea is analyzed to describe the motion of structures within the OC. The two-degree-of freedom model is solved by using the finite difference method and Cramer's rule to investigate and analysis the response of the cochlea. This model is developed to show how an increase of mechanical properties of the CP affects the cochlea response to pure tone excitation. For this, an increase of stiffness is loaded to the BM in order to show its effect on the displacement of the BM and TM. The larger increase stiffness on the BM is applied, the smaller maximum displacement of the coupled response of the cochlea is obtained. Additionally, the deviation of the location of the CF is observed, this change occurs with noticeable changing in the CF distribution of the cochlea at low frequencies. The results obtained in this study suggest that the effect of an increase stiffness of the BM might be associated to some fluctuating hearing loss in the case of low frequencies stimulus.

7 Conflicts of Interest

The authors declare that there are no conflicts of interest regarding the publication of this paper.

References

1. Zwislocki, J.J.: Theory of the acoustical action of the cochlea. Acoust. Soc. Am. **22**, 778–784 (1950). https://doi.org/10.1121/1.1906689
2. Allen, J.B.: Cochlear micromechanics a mechanism for transforming mechanical to neural tuning within the cochlea. Acoust. Soc. Am. **62**(4), 930–939 (1977). https://doi.org/10.1121/1.381586
3. Neely, S.T.: Finite difference solution of a two dimensional mathematical model of the cochlea. Acoust. Soc. Am. **69**, 1386–1393 (1981). https://doi.org/10.1121/1.385820
4. Zwislocki, J.J.: Five decades of research on cochlear mechanics. J. Acoust. Soc. Am. **67**(5), 1679–1685 (1980). https://doi.org/10.1121/1.384294
5. Zwislocki, J.J.: Analysis of cochlear mechanics. Hear. Res. **22**, 155–169 (1986). https://doi.org/10.1016/0378-5955(86)90091-2
6. Cormack, J., Liu, Y., Nam, J.H., Gracewski, S.M.: Two-compartment passive frequency domain cochlea model allowing independent fluid coupling to the tectorial and basilar membrane. J. Acoust. Soc. Am. **37**(3), 1117–1125 (2015). https://doi.org/10.1121/1.4908214

7. Allen, J.B.: Cochlear micromechanics a physical model of transduction. J. Acoust. Soc. Am. **68**(6), 1660–1670 (1980). https://doi.org/10.1121/1.385198
8. Neely, S.T., Kim, D.O.: A model for active elements in cochlear biomechanics. J. Acoust. Soc. Am. **79**(5), 1472–1480 (1986). https://doi.org/10.1121/1.393674
9. Neely, S.T.: A model of cochlear mechanics with outer hair cell motility. J. Acoust. Soc. Am. **94**(1), 137–146 (1993). https://doi.org/10.1121/1.407091
10. Ku, E.M.: Modelling the human cochlea. Doctoral thesis, University of Southampton (UK) (2008). http://eprints.soton.ac.uk/id/eprint/64535
11. Ku, E.M., Elliot, S.J., Lineton, B.: Limit cycle oscillations in a nonlinear state space model of the human cochlea. J. Acoust. Soc. Am. **126**(2), 739–750 (2009). https://doi.org/10.1121/1.3158861
12. Von Békésy, G., Wever, E.G.: Experiments in Hearing. Trans. (McGraw-Hill, NewYork). Chap. 13. pp. 535–634 (1960)
13. Berlin, C.H., Bobbin, R.P.: Hair Cells Micromechanics and Hearing. Singular Thomson Learning (2001)
14. Berlin, C.H., Bobbin, R.P.: Hair Cells Micromechanics and Otoacustic Emission. Singular Thomson Learning (2002)
15. Zum Gottesberge, A.M.M., Gross, O., Becker-Lendzian, U., Massing, T., Vogel, W.F.: Inner ear defects and hearing loss in mice lacking the collagen receptor DDR1. Lab. Investig. **88**(1), 27–37 (2008)
16. Aboulkhouatem, F.E., Kouilily, F., El Khasmi, M., Achtaich, N., Yousfi, N.: The active model: the effect of stiffness on the maximum amplitude displacement of the basilar membrane. Br. J. Math. Comput. Sci. **20**(6), 1–11 (2017). https://doi.org/10.9734/BJMCS/2017/30856
17. Kouilily, F., Aboulkhouatem, F.E., Yousfi, N., Achtaich, N.: Predicting the effect of physical parameters on the amplitude of the passive cochlear model. Rev. Mex. Ing. Biomédica. **39**(1), 105–112 (2018). https://doi.org/10.17488/RMIB.39.1.0
18. Alves, F.R., Ribeiro, F.D.A.Q.: Revision about hearing loss in the Alport's syndrome, analyzing the clinical, genetic and bio-molecular aspects. Revista Brasileira de Otorrinolaringologia. **71**(6), 813–819 (2005)
19. Zehnder, A.F., Adams, J.C., Santi, P.A., et al.: Distribution of type IV collagen in the cochlea in Alport syndrome. Arch Otolaryngol Head Neck Surg. **131**(11), 1007–1013 (2005). https://doi.org/10.1001/archotol.131.11.1007
20. Gross, O., Kashtan, C.E., Rheault, M.N., Flinter, F., Savige, J., Miner, J.H., Perin, L.: Advances and unmet needs in genetic, basic and clinical science in Alport syndrome: report from the 2015 International Workshop on Alport Syndrome. Nephrol. Dial. Transplant. **32**(6), 916–924 (2016). https://doi.org/10.1093/ndt/gfw095
21. Lee, S., Koike, T.: Simulation of the basilar membrane vibration of endolymphatic hydrops. Procedia IUTAM. **24**, 64–71 (2017). https://doi.org/10.1016/j.piutam.2017.08.043
22. Aboulkhouatem, F.E., Kouilily, F., El Khasmi, M., Achtaich, N., Yousfi, N.: The influence of fluid pressure in macromechanical cochlear model. Aust. J. Math. Anal. Appl. **16**(1), 1–9 (2019). https://ajmaa.org/searchroot/files/pdf/v16n1/v16i1p8.pdf
23. De Boer, E.: Mechanics of the cochlea: modelling effects. The Cochlea, pp. 258–317. Springer, New York, NY (1996)

New Variant of the GOST Digital Signature Protocol

Leila Zahhafi and Omar Khadir

Abstract In this paper we propose a new variant of GOST R 34.10-2012 digital signature algorithm. We modified the signature equation to make it more secure against current attacks. We analyze security and complexity of the proposed protocol.

1 Introduction

Since the invention of the public key cryptography in 1979, several methods to cipher secret messages, identify an entity and sign documents were suggested.

Digital signature is a tool that allows to valid identity of the signer and the integrity of the signed document. The process of a digital signature in the public key cryptography starts by the key production. In this step the signer generates the set of his secret keys and calculates his public keys. To sign a given document, the signer uses his secret keys, an encryption function f_e and a hash function h. Then, the verifier can check the validity of the received signature using a decryption function f_d and the set of public parameters of the signer.

As all protocols in the public key cryptography, digital signatures are based on hard mathematical problems as factorization of large composite integers and the discrete logarithm problem.

In 1985, Koblitz [5] and Miller [8] invented the elliptic curves cryptosystems. The older discrete logarithm problem DLP was replaced by the elliptic curves discrete logarithm problem ECDLP that is considered as more difficult to be computed.

The GOST R 34.10-2012 [3] is one of the Russian cryptographic standard algorithms, it is based on elliptic curve operations. Its security is related to the complexity to solve an elliptic curve discrete logarithm problem and the efficiency of the hash function used [4]. Several variants of the GOST algorithm were proposed as

L. Zahhafi (✉) · O. Khadir
Laboratory of Mathematics, Cryptography, Mechanics and Numerical Analysis (LMCMNA), University Hassan II of Casablanca, Casablanca, Morocco
e-mail: leila.zahhafi@gmail.com

S. Melliani et al. (eds.), *Applied Mathematics and Modelling in Finance, Marketing and Economics*, Studies in Computational Intelligence 1114,
https://doi.org/10.1007/978-3-031-42847-0_8

GOST-I and GOST-II suggested by Trieu Quang Phong and Nguyen Quoc Toan [9] in previous edition of the Current Trends in Cryptology workshop (CTCrypt'2017).

In this work, We present an amelioration of The GOST R 34.10-2012 signature algorithm. We propose a new variant in which we modified the basic signature equation by adding a new variable. We show how our update reenforces the security of the signature scheme.

The paper is organized as follow: We start by a short description of the elliptic curves theory. The third section recalls the original algorithm of GOST R 34.10-2012. We present in section four our main contribution and we finish by a conclusion in section five.

2 Review of Elliptic Curves Theory

In this section, we present the basic theory of elliptic curves.

Let p be a large prime integer, we define the elliptic curve E over a finite filed F_p with a characteristic different of 2 and 3 in order to write E in the simplified Weierstrass form.

The equation of the elliptic curve E is as follow:

$$y^2 \equiv x^3 + ax + b \ \ mod\ p \tag{1}$$

where a and b are two integers smaller than p.

2.1 Group Law

Let $E(F_p)$ be an elliptic curve defined by Eq. (1). We describe the group low on $E(F_p)$ [10, p. 12] as follow:

- $\mathcal{O}$ Point at infinity of the curve $E(F_p)$: $(x_1, y_1) + \mathcal{O} = \mathcal{O} + (x_1, y_1) = (x_1, y_1)$ with $(x_1, y_1) \in E(F_p)$.
- The opposite (x_1, y_1) is $-(x_1, y_1) = (x_1, -y_1)$.

The following formula presents addition of two points $P = (x_1, y_1)$ and $Q = (x_2, y_2)$ in $E(F_p)$ with $P \neq -Q$ (else, we get Q is the opposite of P then $P + Q = \mathcal{O}$).

Let $S = (x_3, y_3)$ be the coordinates of the point $P + Q$.

1. If $x_1 \neq x_2$ then:
   $x_3 = m^2 - x_1 - x_2$ and $y_3 = m(x_1 - x_2) - y_1$. With: $m = \frac{y_2 - y_1}{x_2 - x_1}$.
2. If $x_1 = x_2$ and $y_1 \neq y_2$ then $S = \mathcal{O}$.

3. If $P = Q$ and $y_1 \neq 0$ then:
   $x_3 = m^2 - 2x_1$ et $y_3 = m(x_1 - x_3) - y_1$. Such that: $m = \frac{3x_1^2+a_4}{2y_1}$.
4. If $P = Q$ and $y_1 = 0$ then $S = \mathcal{O}$.

Theorem 1 *The set of points of the elliptic curve $E(F_p)$ forms an Abelian group whose identity element is the point at infinity $\mathcal{O}$ [10, p. 15].*

2.2 Multiplication of a Point by an Integer

Lets $E(F_p)$ be a elliptic curve, $P \in E(F_p)$, and an integer $k \in N^*$.

We define the multiple of P by k, $kP = \underbrace{P + \cdots + P}_{k\ times}$.

If $k = 0$, we define $kP = \mathcal{O}$.

2.3 The Discrete Logarithm Problem

Let F_p be a group, p a large prime and $\alpha \in F_p$. The discrete logarithm problem in F_p is to solve the following equation:

$$a \equiv b^{\alpha}\ mod\ p \tag{2}$$

With α is the unknown variable.

In the case where $F_p = E(F_p)$ an elliptic curve, the discrete logarithm problem is to find x that verifies:

$$\mathscr{A} = x\mathscr{B} \tag{3}$$

Such that $\mathscr{A}$ and $\mathscr{B}$ are points in $E(F_p)$.

3 GOST R 34.10-2012 Digital Signature Scheme [3]

3.1 Keys Production

The signer, Alice starts by choosing a prime integer p and an elliptic curve $E(F_p)$ defined by the following equation:

$$y^2 \equiv x^2 + ax + b\ mod\ p \tag{4}$$

We define the integer m as an elliptic curve $E(F_p)$ points group order and the prime integer q such that: $m = nq$ with $2^{254} < q < 2^{256}$ or $2^{508} < q < 2^{512}$.

We select the point $\mathscr{P} = (x_P, y_P) \neq \mathscr{O}$ of the curve $E(F_p)$, with $q\mathscr{P} = \mathscr{O}$.

Then, we fix a secure hash function h [4] which maps messages onto binary vectors of an l-bit length with $l = 256$ if $2^{254} < q < 2^{256}$ and $l = 512$ if $2^{508} < q < 2^{512}$.

Alice chooses her secret signature key d as a positive integer smaller than q. She calculates the point $\mathscr{Q} = (x_Q, y_Q)$ such that $\mathscr{Q} = d\mathscr{P}$.

3.2 The Signature Generation

To sign a message M, Alice have to execute the following algorithm:

Algorithm 1 The GOST R 34.10-2012 signature algorithm

Require: : $d, M, h(.), q, E(F_p)$.
Step 1:
$H \leftarrow h(M)$;
Step 2:
Determine the random integer α, with H is the binary representation of α;
$e \leftarrow \alpha \ mod \ q$;
if $e = 0$ **then**
  $e \leftarrow 1$;
end if
Step 3:
Select the positive integer k with $k < q$;
Step 4:
$\mathscr{C} \leftarrow k\mathscr{P}$;
$r \leftarrow x_C \ mod \ q$;
if $r = 0$ **then**
  return to step 3;
end if
Step 5:
$s \leftarrow (rd + ke) \ mod \ q$;
if $s = 0$ **then**
  return to step 3;
end if
Step 6:
Determine the binary vectors R and S, corresponding to r and s;
$\zeta \leftarrow (R||S)$
End

3.3 Signature Verification

The verifier, Bob downloads the signature ζ sent by the signer. He can check the validity of the received data using the following algorithm:

Algorithm 2 The GOST R 34.10-2012 signature algorithm

Require: : $d, M, h(.), q, E(F_p)$.
Step 1:
$H \leftarrow h(M)$;
Step 2:
Determine the random integer α, with H is the binary representation of α;
$e \leftarrow \alpha \; mod \; q$;
if $e = 0$ **then**
$e \leftarrow 1$;
end if
Step 3:
Select the positive integer k with $k < q$;
Step 4:
$\mathscr{C} \leftarrow k\mathscr{P}$;
$r \leftarrow x_C \; mod \; q$;
if $r = 0$ **then**
return to step 3;
end if
Step 5:
$s \leftarrow (rd + ke) \; mod \; q$;
if $s = 0$ **then**
return to step 3;
end if
Step 6:
Determine the binary vectors R and S, corresponding to r and s;
$\zeta \leftarrow (R||S)$
End

3.4 Signature Verification

The verifier, Bob downloads the signature ζ sent by the signer. He can check the validity of the received data using the following algorithm:

Algorithm 3 The GOST R 34.10-2012 signature algorithm

Require: : ζ, $\mathscr{Q}$, M, $h(.)$, q, $E(F_p)$.
Step 1:
Determine integers r and s using the received signature ζ;
if $0 < r < q$ and $0 < s < q$ **then**
Go to the next step;
else
Print ("Invalid Signature");
end if
Step 2:
$H \leftarrow h(M)$;
Step 3:
Determine the random integer α, with H is the binary representation of α;
$e \leftarrow \alpha \; mod \; q$;
if $e = 0$ **then**
$e \leftarrow 1$;
end if
Step 4:
$v \leftarrow e^{-1} \; mod \; q$;
Step 5:
$z_1 = sv \; mod \; q$;
$z_2 = -rv \; mod \; q$;
Step 6:
$\mathscr{C} \leftarrow z_1\mathscr{P} + z_2\mathscr{Q}$;
$R \leftarrow x_C \; mod \; q$;
Step 7:
if $R = r$ **then**
Print ("Valid signature");
else
Print ("Invalid signature");
end if
End

4 Our Contribution

In this section, we present our contribution. We adopt the same notations as is in Ref. [3].

4.1 Keys Production

Let $E_p(a, b)$ be the elliptic curve defined by the equation:

$$y^2 \equiv x^3 + ax + b \; mod \; p \tag{5}$$

where p is a large prime number. We generate all signature parameters as in Sect. 3.1.

Alice public key are $(a, b, p, q, \mathscr{P}, \mathscr{Q})$ and her private signature key is the integer d.

4.2 The Signature Equation

To sign a message M whose hash is $H = h(M)$ and $e = \alpha \ mod \ q$ with H is the binary representation of the integer α, Alice must solve the equation:

$$s\mathcal{P} = x_C\mathcal{Q} + x_{C'}\mathcal{C}(x_C, y_C) + e\mathcal{C}'(x_{C'}, y_{C'}) \tag{6}$$

where the unknown variables x_C, y_C, $x_{C'}$, and $y_{C'}$ are in $\{1, 2, \ldots, p-1\}$ and s in $\{1, 2, \ldots, q-1\}$.

4.3 How Can Alice Generate a Signature

We describe in the following the signature generation algorithm of our proposed variant.

Algorithm 4 The variant of GOST R 34.10-2012 signature algorithm

Require: : $d, M, h(.), q, E(F_p)$.
Step 1:
$H \leftarrow h(M)$;
Step 2:
Determine the random integer α, with H is the binary representation of α;
$e \leftarrow \alpha \ mod \ q$;
if $e = 0$ **then**
$e \leftarrow 1$;
end if
Step 3:
Select the positive integers k and k' with $k, k' < q$;
Step 4:
$\mathcal{C} \leftarrow k\mathcal{P}$;
$\mathcal{C}' \leftarrow k'\mathcal{P}$;
$r \leftarrow x_C(mod \ q)$;
$r' \leftarrow x_{C'}(mod \ q)$;
if $r = 0$ or $r' = 0$ **then**
return to step 3;
end if
Step 5:
$s \leftarrow (rd + r'k + ek') \ mod \ q$;
if $s = 0$ **then**
return to step 3;
end if
Step 6:
Determine the binary vectors R, R' and S, corresponding to r, r' and s;
$\zeta \leftarrow (R||R'||S, \mathcal{C})$
End

4.4 Signature Verification

The verifier, Bob, downloads the signature ζ sent by the signer Alice. He can check the validity of the digital signature as follow:

4.5 Numerical Example

Let us take the same signature parameters proposed in the original paper of GOST R 34.10-2012 signature [3, p. 15].

Consider the elliptic curve $y^2 \equiv x^3 + ax + b \bmod p$, where
$p = 57896044618658097711785492504343953926634992332820282019728792003956564821041$

Algorithm 5 The variant of GOST R 34.10-2012 signature verification algorithm

Require: : $\zeta, \mathcal{Q}, M, h(.), q, E(F_p)$.
Step 1:
Determine integers $r, r\prime$ and s using the received signature ζ;
if $0 < r < q, 0 < r\prime < q$ and $0 < s < q$ **then**
  Go to the next step;
else
  Print ("Invalid Signature");
end if
Step 2:
$H \leftarrow h(M)$;
Step 3:
Determine the random integer α, with H is the binary representation of α;
$e \leftarrow \alpha \bmod q$;
if $e = 0$ **then**
  $e \leftarrow 1$;
end if
Step 4:
$v \leftarrow e^{-1} \bmod q$;
Step 5:
$z_1 = sv \bmod q$;
$z_2 = -rv \bmod q$;
$z_3 = -r\prime v \bmod q$;
Step 6:
$\mathcal{C}\prime \leftarrow z_1\mathcal{P} + z_2\mathcal{Q} + z_3\mathcal{C}$;
$R\prime \leftarrow x_{C\prime} \bmod q$;
Step 7:
if $R\prime = r\prime$ **then**
  Print ("Valid signature");
else
  Print ("Invalid signature");
end if
End

$a = 7$ and
$b = 43308876546767276905765904595650931995942111794451039583252968$
842033849580414.
The elliptic curve points group order is:
$m = 57896044618658097711785492504343953927082934583725450622380973$
592137631069619.
The order q of cyclic subgroup of elliptic curve points group is:
$q = 57896044618658097711785492504343953927082934583725450622380973$
592137631069619.
The elliptic curve point $\mathscr{P}$ coordinates are:
$x_P = 2$
$y_p = 40189740565390375033354494229370597756357393899055450806909793$
65213431566280.
The secret signature key is:
$d = 55441196065363246126355624130324183196576709222340016572108097$
750006097525544.
The verification key $\mathscr{Q}$ Coordinates are:
$x_Q = 57520216126176808443631405023338071176630104906313632182896741$
342206604859403
$y_Q = 17614944419213781543809391949654080031942662045363639260709847$
859438286763994.
As in [3], we suppose that:
$e = 20798893674476452017134061561508270130637142515379653289952617$
252661468872421
and
$k = 53854137677348463731403841147996619241504003434302020712960838$
528893196233395.
The point $\mathscr{C} = k\mathscr{P}$ has the following coordinates:
$x_C = 29700980915817952874371204983938256990422752107994319651632687$
982059210933395
$y_C = 32842535278684663477094665322517084506804721032454543268132854$
556539274060910.
We select as random:
$k\prime = 48836620439032517596916187958502728852845976338323070089561057$
540896822659464.
We get the point $\mathscr{C}\prime = k\prime\mathscr{P}$ with the following coordinates:
$x_{C\prime} = 42885619182756687869977362675168212010395930385700877385308454$
978145277505116
$y_{C\prime} = 25177044364870406864620533587369770559579786350743723259891731$
834739002200259.
Integers $r = x_C \ mod \ q$ and $r\prime = x_{C\prime} \ mod \ q$ take values:
$r = 29700980915817952874371204983938256990422752107994319651632687$
982059210933395
$r\prime = 42885619182756687869977362675168212010395930385700877385308454$
978145277505116.

We obtain the parameter $s = (rd + r\prime k + ek\prime) \ mod \ q$:
$s = 5649171002854384536917142657507489548678986646810029822222539372101954045\,1658$.

We verify the above signature using Algorithm 5.
We calculate the parameter $v = e^{-1} \ mod \ q$, so:
$v = 176866836059344686773017138249002685627468830806754967152880365724311457\,18978$.
We compute: $z_1 = sv \ mod \ q = 14124443569344013609788848698488144848745703337391824853309123521165596200688$
$z_2 = -rv \quad mod \quad q = 14171998427343472112515917969500765769246655838972862114499932653333671\,09221$
$z_3 = -r\prime v \quad mod \quad q = 99959937513850028119087043798199750998486669442857384504342966938012063\,71380$.
We get the point $\mathscr{C}\prime = z_1\mathscr{P} + z_2\mathscr{Q} + z_3\mathscr{C}$ with the following coordinates:
$x_{C\prime} = 4288561918275668786997736267516821201039593038570087738530845497814527\,7505116$
$y_{C\prime} = 2517704436487040686462053358736977055957978635074372325989173183473900\,2200259$.
Then we find the parameter $R\prime = x\prime_C \ mod \ q = 42885619182756687869977362675168212010395930385700877385308454978145277505116$.
We have $R\prime = r\prime$. Conclusion: the signature is valid.

4.6 Security Analysis

Suppose That Oscar is an attacker. We describe in the following some possible ways that Oscar can use to impersonate Alice by signing a given message M.

Attack 1: Knowing the public parameters $(p, q, \mathscr{P}, \mathscr{Q})$, Oscar is not able to determine the secret key d of Alice as he is confronted to the discrete logarithm problem to solve the equation: $\mathscr{Q} = x\mathscr{P}$ which means running $\sqrt{\log p \log \log p}$ operations [1].

Attack 2: Assume that Oscar fixes arbitrary two parameters and tries to find the third one using Eq. 6.

1. If he fixes $\mathscr{C}\prime = (x_{C\prime}, y_{C\prime})$ and $\mathscr{C} = (x_C, y_C)$, then he will be confronted to the discrete logarithm problem: $A = sP$ with: $\mathscr{A} = x_C\mathscr{Q} + x_{C\prime}\mathscr{C} + e\mathscr{C}\prime$.
2. If he fixes $\mathscr{C}$ and s and tries to find the elliptic curve point $\mathscr{C}\prime$, then he will be confronted to the equation: $A = x_{C\prime}\mathscr{C} + e\mathscr{C}\prime$, with $A = s\mathscr{P} - x_C\mathscr{Q}$.
3. If Oscar fixes $\mathscr{C}\prime$ and s and tries to find the parameter $\mathscr{C}$, then he will be confronted to the equation: $A = x_C\mathscr{Q} + x_{C\prime}\mathscr{C}$ with $A = s\mathscr{P} - e\mathscr{C}\prime$.

Attack 3: If Oscar intercepts n valid signatures $\zeta_i = (R_i||R\prime_i||S_i, \mathscr{C}_i)$ for n different messages where $\forall i \in \{1, 2, \ldots, n\}$ and n is a natural integer, then he constructs the following system:

$$(S)\begin{cases} s_1 = (r_1 d + r\prime_1 k_1 + e_1 k\prime_1) \; mod \; q \\ s_2 = (r_2 d + r\prime_2 k_2 + e_2 k\prime_2) \; mod \; q \\ \quad \vdots \quad \vdots \quad \vdots \\ s_n = (r_n d + r\prime_n k_n + e_n k\prime_n) \; mod \; q \end{cases}$$

Using the system (S) Oscar can propose a valid solution for the unknown variables $r_i, r\prime_i, s_i$ and d. Since the secret key of Alice d has an unique value, Oscar will never be able to determine what possibility of d is the right one.

Attack 4: Assume that the signature scheme is used without hash functions. Oscar puts: $\mathscr{C}(x_C, y_C) = k_1\mathscr{P} + k_2\mathscr{Q}$ and $\mathscr{C}\prime(x_{C\prime}, y_{C\prime}) = k\prime_1\mathscr{P} + k\prime_2\mathscr{Q}$ with $k_1, k_2, k\prime_1, k\prime_2 < q - 1$.

Then he can obtain:

$$(S') = \begin{cases} M = \frac{s - r\prime k_1}{k\prime_1} \; mod \; q \\ s = r\prime k_1 - \frac{k\prime_1}{k\prime_2}(r + r\prime k_2) \; mod \; q \end{cases}$$

Note that $\zeta = (R||R\prime||S, \mathscr{C})$, with R, $R\prime$ and S are the binary vectors of r, $r\prime$ and s, is a valid signature for a message M but this attack is not realistic.

4.7 *Running Time*

Let T_{Mul}, T_{Sq}, T_{PM}, T_{PA}, T_{PD} and T_h be times to calculate respectively a multiplication, squaring, point multiplication, point addition, doubling of a point and a hush function.

We summarize in the table below numbers of executed operations in the proposed signature protocol:

	T_{PM}	T_{PA}	T_{Mul}	T_h
Keys production	1	0	0	0
The signature generation	2	0	3	1
The signature verification	3	2	3	1

To compute a point multiplication: nQ where n is an integer and Q a point, we need to execute $\log(n)$ point additions and doubling.

We compute 9 multiplications and 1 squaring for a point addition, and 3 multiplications and 4 squaring for a point doubling [2, 6].

We assume that $T_{Sq} = O((\log n)^3)$ and $T_{Mul} = O((\log n)^2)$ [7, p. 72]. The total complexity T_{tot} of the signature protocol is:
$T_{tot} = 6T_{PM} + 2T_{PA} + 6T_{Mul} + 2T_h = O((log(q))^2 + (log(q))^3) + 2T_h$.

5 Conclusion

In this article, we worked on the GOST R 34.10-2012 digital signature. We ameliorated the original signature algorithm to make it more secure. We analyzed security and complexity of our signature protocol.

References

1. Adleman, L.M., Pomerance, C., Rumely, R.S.: On distinguishing prime numbers from composite numbers. Ann. Math. 173–206 (1983)
2. Bernstein, D.J., Lange, T.: Faster addition and doubling on elliptic curves. In: Kurosawa, K. (ed.) Advances in Cryptology ASIACRYPT. ASIACRYPT Lecture Notes in Computer Science, vol. 4833. Springer, Berlin, Heidelberg (2007)
3. Dolmatov, V., Degtyarev, A.: GOST R 34.10−2012: Digital Signature Algorithm, pp. 1–21. Cryptocom, Ltd. (2013)
4. Dolmatov, V., Degtyarev, A.: GOST R 34.11−2012: Hash Function, pp. 1–40. Cryptocom, Ltd. (2013)
5. Koblitz, N.: Elliptic curve cryptosystems. Math. Comput. **48**, 203–209 (1987)
6. Koblitz, N.: Algebraic Aspects of Cryptography. Springer, Berlin (1999)
7. Menezes, A.J., van Oorschot, P.C., Vanstone, S.A.: Handbook of Applied Cryptography (1996)
8. Miller, V.: Uses of elliptic curves in cryptography. Advances in Cryptology-Crypto'85. Lecture Notes in Computer Science, vol. 218, pp. 417–426. Springer (1986)
9. Trieu, Q.P., Nguyen, Q.T.: Some Security Comparisons of GOST R 34.10-2012 and ECDSA Signature Schemes. CTCrypt (2017)
10. Washington, L.C.: Elliptic Curves: Number Theory and Cryptography, 2nd edn. (2008)

Existence and Uniqueness Solutions of Fuzzy Fractional Integration-Differential Problem Under Caputo gH-Differentiability

S. Melliani, E. Arhrrabi, M. Elomari, and L. S. Chadli

Abstract This paper is devoted to considering the local existence and uniqueness of fuzzy fractional integration-differential problem under Caputo-type fuzzy fractional derivative employing the contraction principle. Some patterns are presented to describe these results.

1 Introduction

The theory of fractional calculus, which deals with the investigation and applications of derivatives and integrals of arbitrary order has a long history. The theory of fractional calculus developed mainly as a pure theoretical field of mathematics, in the last decades it has been used in various fields as rheology, viscoelasticity, electrochemistry, diffusion processes, etc. [32, 33]. Fractional calculus have undergone expanded study in recent years as a considerable interest both in mathematics and in applications. One of the recently influential works on the subject of fractional calculus is the monograph of Podlubny [49] and the other is the monograph of Kilbas et al. [33]. The fractional differential equations have great application potential in modeling a variety of real world physical problems, which deserves further investigations. Among these we might include the modeling of earthquakes, the fluid dynamic traffic model with fractional derivatives, the measurement of viscoelastic material properties, etc. Consequently, several research papers were done to investigate the theory and solutions of fractional differential equations (see [18, 21, 35, 37] and references therein).

The concept of solution for fractional differential equations with uncertainty was introduced by Agarwal, Lakshmikantham and Nieto [1]. They considered Riemann–Liouville differentiability concept based on the Hukuhara differentiability to solve fuzzy fractional differential equations. Arshad and Lupulescu [12] proved some

S. Melliani (✉) · E. Arhrrabi · M. Elomari · L. S. Chadli
LMACS, Laboratory of Applied Mathematics and Scientific Calculus, Sultan Moulay Sliman University, Beni Mellal, Morocco
e-mail: saidmelliani@gmail.com

S. Melliani et al. (eds.), *Applied Mathematics and Modelling in Finance, Marketing and Economics*, Studies in Computational Intelligence 1114,
https://doi.org/10.1007/978-3-031-42847-0_9

results on the existence and uniqueness of solution to fuzzy fractional differential equation under Hukuhara fractional Riemann–Liouville differentiability. Some existence results for nonlinear fuzzy differential equations of fractional order involving the Riemann–Liouville derivative have been proposed in [30]. The solutions of fuzzy fractional differential equations are investigated by using the fuzzy Laplace transforms in [51]. Recently, the concepts of fractional derivatives for a fuzzy function are either based on the notion of Hukuhara derivative [25] or on the notion of strongly generalized derivative. The concept of Hukuhara derivative is old and well known, but the concept of strongly generalized derivative was recently introduced by Bede and Gal [13]. Using this new concept of derivative, the classes of fuzzy differential equations have been extend and studied in some papers such as: Ahmad et al. [4], Allahviranloo et al. [9–11], Bede et al. [14–17], Gasilov [20], Khastan et al. [27–29], Malinowski [41–43] and Nieto [46]. Furthermore, by using this new concept of derivative, Allahviranloo et al. [7, 8] have studied the concepts about generalized Hukuhara fractional Riemann–Liouville and Caputo differentiability of fuzzy-valued functions. Later, authors have proved the existence and uniqueness of solution for fuzzy fractional differential equation by using different methods. Alikhani et al. [6] have proved the existence and uniqueness results for nonlinear fuzzy fractional integral and integrodifferential equations by using the method of upper and lower solutions. Mazandarani et al. [44] studied the solution to fuzzy fractional initial value problem under Caputo-type fuzzy fractional derivatives by a modified fractional Euler method. Besides, authors studied some results on the existence and uniqueness of solution to fuzzy fractional differential equation under Caputo type-2 fuzzy fractional derivative and the definition of Laplace transform of type-2 fuzzy number-valued functions [45]. Salahshour et al. [50] proposed some new results toward existence and uniqueness of solution of fuzzy fractional differential equation. According to the concept of Caputo-type fuzzy fractional derivative in the sense of the generalized fuzzy differentiability, Fard et al. [19] extended and established some definitions on fuzzy fractional calculus of variation and provide some necessary conditions to obtain the fuzzy fractional Euler–Lagrange equation for both constrained and unconstrained fuzzy fractional variational problems. Ahmad et al. [5] proposed a new interpretation of fuzzy fractional differential equations and present their solutions analytically and numerically. The proposed idea is a generalization of the interpretation given in [3, 4], where the authors used Zadeh's extension principle to interpret fuzzy differential equations.

In real world systems, delays can be recognised everywhere and there has been widespread interest in the study of delay differential equations for many years. Therefore, delay differential equations (or, as they are called, functional differential equations) play an important role in an increasing number of system models in biology, engineering, physics and other sciences. There exists an extensive amount of literature dealing with delay differential equations and their applications; the reader is referred to the monographs [22, 34], and the references therein. The study of fuzzy delay differential equations is expanding as a new branch of fuzzy mathematics. Both theory and applications have been actively discussed over the last few years. In the literature, the study of fuzzy delay differential equations has several interpretations.

The first one is based on the notion of Hukuhara derivative. Under this interpretation, Lupulescu established the local and global existence and uniqueness results for fuzzy delay differential equations. The second interpretation was suggested by Khastan et al. [29] and Hoa et al. [24]. In this setting, Khastan et al. proved the existence of two fuzzy solutions for fuzzy delay differential equations using the concept of generalized differentiability. Hoa et al. established the global existence and uniqueness results for fuzzy delay differential equations using the concept of generalized differentiability. Moreover, authors have extended and generalized some comparison theorems and stability theorem for fuzzy delays differential equations with definition a new Lyapunov-like function. Besides that, some very important extensions of the fuzzy delay differential equations are the fuzzy delay integro-differential equations and the random fuzzy delay differential equations [23, 52–54]. Combining the two aspects introduced, fractional calculus and fuzzy delay differential equations, we get fuzzy fractional delay differential equations. Furthermore, fuzzy fractional delay differential equations are a very recent topic.

This paper is organized as follows: In Sect. 2, we present the basic notations of the Riemann–Liouville fractional integral and Caputo fractional derivative for fuzzy functions. In Sect. 3, we study the existence and uniqueness theorems of solutions for two general forms of fuzzy fractional functional integral differential equations by the contraction principle. Some examples of this class having two different solutions were presented in Sect. 4.

2 Preliminaries

The basic definition of fuzzy numbers is mentioned in [36]. Let E denote the set of fuzzy subsets of the real axis, if $\omega : \mathbb{R} \longrightarrow [0, 1]$, satisfying the following properties:

(i) ω is normal, that is, there exists $z_0 \in \mathbb{R}$ such that $\omega(z_0) = 1$,
(ii) ω is fuzzy convex, that is, for $0 \leq \lambda \leq 1$

$$\omega(\lambda z_1 + (1 - \lambda)z_2) \geq min\{\omega(z_1), \omega(z_2)\}, \quad for\ any\ z_1, z_2 \in \mathbb{R},$$

(iii) ω is upper semicontinuous on $\mathbb{R}$,
(iv) $[\omega]^0 = cl\{z \in \mathbb{R} : \omega(z) > 0\}$ is compact, where cl denotes the closure in $(\mathbb{R}, |\cdot|)$.

Then E is called the space of fuzzy number. For $r \in (0, 1]$, we denote $[\omega]^r = \{z \in \mathbb{R} \mid \omega(z) \geq r\}$ and $[\omega]^0 = \{z \in \mathbb{R} \mid \omega(z) > 0\}$. From the conditions (i) to (iv), it follows that the $r - level$ set of ω, $[\omega]^r$, is a nonempty compact interval, for all $r \in [0, 1]$ and any $\omega \in E$.

The notation $[\omega]^r = [\underline{\omega}(r), \overline{\omega}(r)]$, denotes explicitly the $r - level$ set of ω, for $r \in [0, 1]$. We refer to $\underline{\omega}$ and $\overline{\omega}$ as the lower and upper branches of ω, respectively. For $\omega \in E$, we define the length of the $r - level$ set of ω as $len([\omega]^r) = \overline{\omega}(r) - \underline{\omega}(r)$.

For addition and scalar multiplication in fuzzy set space E, we have $[\omega_1 + \omega_2]^r = [\omega_1]^r + [\omega_2]^r$, $[\lambda\omega]^r = \lambda[\omega]^r$. Moreover, the multiplication $\omega_1\omega_2$ are defined by

$$[\omega_1\omega_2]^r = \left[min\{\underline{\omega}_1(r)\underline{\omega}_2(r), \underline{\omega}_1(r)\overline{\omega}_2(r), \overline{\omega}_1(r)\underline{\omega}_2(r), \overline{\omega}_1(r)\overline{\omega}_2(r)\}, max\{\underline{\omega}_1(r)\underline{\omega}_2(r), \underline{\omega}_1(r)\overline{\omega}_2(r), \overline{\omega}_1(r)\underline{\omega}_2(r), \overline{\omega}_1(r)\overline{\omega}_2(r)\}\right]$$

The Hausdorff distance between fuzzy numbers is given by

$$D_0[\omega_1, \omega_2] = \sup_{0\le r\le 1} \{\mid \underline{\omega}_1(r) - \underline{\omega}_2(r) \mid, \mid \overline{\omega}_1(r) - \overline{\omega}_2(r) \mid\}.$$

The metric space (E, D_0) is complete metric space and the following properties of the metric D_0 are valid (see [36]).

$$D_0[\omega_1 + \omega_3, \omega_2 + \omega_3] = D_0[\omega_1, \omega_2],$$

$$D_0[\lambda\omega_1, \lambda\omega_2] =\mid \lambda \mid D_0[\omega_1, \omega_2],$$

$$D_0[\omega_1, \omega_2] \le D_0[\omega_1, \omega_3] + D_0[\omega_3, \omega_2],$$

for all $\omega_1, \omega_2, \omega_3 \in E$ and $\lambda \in \mathbb{R}$. Let $\omega_1, \omega_2 \in E$, if there exists $\omega_3 \in E$ such that $\omega_1 = \omega_2 + \omega_3$ then ω_3 is called the H-difference of ω_1, ω_2. We denote the ω_3 by $\omega_1 \ominus \omega_2$. Let us remark that $\omega_1 \ominus \omega_2 \neq \omega_1 + (-1)\omega_2$.

Definition 1 ([17]) The generalized Hukuhara difference of two fuzzy numbers $\omega_1, \omega_2 \in E$ (gH-difference for short) is defined as follows:

$$\omega_1 \ominus_{gH} \omega_2 = \omega_3 \Leftrightarrow \begin{cases} (i)\omega_1 = \omega_2 + \omega_3, \\ or(ii)\omega_2 = \omega_1 + (-1)\omega_3. \end{cases}$$

The generalized Hukuhara differentiability was introduced in [17].

Definition 2 Let $t \in (a, b)$ and h such that $t + h \in (a, b)$, then the generalized Hukuhara derivative of fuzzy-valued function $x : (a, b) \longrightarrow E$ at t is defined as

$$D_{gH}x(t) = \lim_{h\longrightarrow 0} \frac{x(t+h) \ominus_{gH} x(t)}{h}. \tag{1}$$

If $D_{gH}x(t) \in E$ satisfying (1) exists, we say that x is generalized Hukuhara differentiable (gH-differentiable for short) at t. Also, we say that x is $[(i) - gH]$-differentiable at t if (i) $[D_{gH}x(t)]^r = [\underline{x}'(t, r), \overline{x}'(t, r)]$, and that x is $[(ii) - gH]$-differentiable at t if (ii) $[D_{gH}x(t)]^r = [\overline{x}'(t, r), \underline{x}'(t, r)]$, $r \in [0, 1]$.

Theorem 1 ([17]) *Let $x : [a, b] \longrightarrow E$ be such that $[x(t)]^r = [\underline{x}'(t, r), \overline{x}'(t, r)]$ for $t \in [a, b]$, $r \in [0, 1]$. If the real-valued function $\underline{x}(t, r)$ and $\overline{x}(t, r)$ are differentiable at $t \in [a, b]$, then the function x is gH-differentiable at $t \in [a, b]$ and*

$$[D_{gH}x(t)]^r = \left[min\{\underline{x}'(t,r), \overline{x}'(t,r)\}, max\{\underline{x}'(t,r), \overline{x}'(t,r)\}\right]. \tag{2}$$

Let $I = [a, b] \subset \mathbb{R}$ be a compact interval we shall use the notation $C([a, b], E) = \{x : I \longrightarrow E \mid x \ is \ continuous\}$, where the continuous is one-sided at endpoints a, b. In the space $C([a, b], E)$, we consider the following metric:

$$D_0^*[x, z] = \sup_{t \in [a,b]} D_0[x(t), z(t)].$$

It is known that $(C([a, b], E), D_0^*)$ is a complete metric space. Also, we denote the space of all Lebesque integrable fuzzy-valued functions on $[a, b]$ by $L([a, b], E)$. Let $x \in C([a, b], E)$, we say that $x \in L(C([a, b], E), D_0^*)$ if and only if $D_0[\int_a^b x(t)dt, \widehat{0}] < \infty$.

Definition 3 The Riemann–Liouville fractional integral operator of order $\alpha > 0$, of a real-valued function $\varphi \in L^1[a, b]$, is defined as

$$I_{a^+}^{\alpha}\varphi(t) = \frac{1}{\Gamma(\alpha)} \int_a^t (t-s)^{\alpha-1}\varphi(s)ds,$$

where $\Gamma(.)$ is the Euler gamma function.

Definition 4 Let $\varphi : [a, b] \longrightarrow \mathbb{R}$, the Caputo fractional derivative of order $\alpha > 0$, $m - 1 < \alpha < m, m \in \mathbb{N}$, is defined as

$${}^C D_{a^+}^{\alpha} = \frac{1}{\Gamma(m-\alpha)} \int_a^t (t-s)^{m-\alpha-1}\varphi^{(m)}(s)ds,$$

where the function $\varphi(t)$ has absolutely continuous derivatives up to order $(m - 1)$. If $\alpha \in (0, 1)$, then

$${}^C D_{a^+}^{\alpha} = \frac{1}{\Gamma(1-\alpha)} \int_a^t (t-s)^{-\alpha}\varphi'(s)ds.$$

Definition 5 ([8]) Let $x : [a, b] \longrightarrow E$, the fuzzy Rieman–Liouville integral of fuzzy-valued function x is defined as follows:

$$(\mathscr{J}_{a^+}^{\alpha}x)(t) = \frac{1}{\Gamma(\alpha)} \int_a^t (t-s)^{\alpha-1}x(s)ds. \tag{3}$$

For $a \leq t$, and $0 < \alpha \leq 1$. For $=\alpha = 1$, we set $\mathscr{J}_a^1 = I$, the identity operator.

The fuzzy gH-fractional Caputo derivative was introduced in [7].

Definition 6 Let $D_{gH} \in C([a, b], E) \cap L([a, b], E)$. The fuzzy gH-fractional Caputo differentiability of fuzzy-valued function x ($[gH]_a^C$-$differentiable\ for\ short$) is defined as following:

$$ {}_{gH}^{C}\mathscr{D}_{a^+}^{\alpha}x(t) = \mathscr{J}_{a^+}^{1-\alpha}(D_{gH}x)(t) = \frac{1}{\Gamma(1-\alpha)}\int_a^t (t-s)^{-\alpha}(D_{gH}x)(s)ds, $$

where $0 < \alpha \leq 1, t > a$.

Lemma 1 ([7]) *Suppose that $x : [a, b] \longrightarrow E$ be a fuzzy function and $D_{gH}x(t) \in C([a, b], E) \cap L([a, b], E)$. Then*

$$ \mathscr{J}_{a^+}^{\alpha}({}_{gH}^{C}\mathscr{D}_{a^+}^{\alpha}x)(t) = x(t) \ominus_{gH} x(a). $$

Theorem 2 ([7]) *Let $D_{gH}x(t) \in C([a, b], E) \cap L([a, b], E)$ be such that $[x(t)]^r = [\underline{x}(t, r), \overline{x}(t, r)]$ for $0 \leq r \leq 1, t \in [a, b]$, then the function x is $[gH]_{\alpha}^{C}$-differentiable at $t \in [a, b]$ and*

$$ [{}_{gH}^{C}\mathscr{D}_{a^+}^{\alpha}x(t)]^r = \left[min\{{}^{C}\mathscr{D}_{a^+}^{\alpha}\underline{x}(t, r), {}^{C}\mathscr{D}_{a^+}^{\alpha}\overline{x}(t, r)\}, max\{{}^{C}\mathscr{D}_{a^+}^{\alpha}\underline{x}(t, r), {}^{C}\mathscr{D}_{a^+}^{\alpha}\overline{x}(t, r)\}\right]. \tag{4} $$

Where ${}^{C}\mathscr{D}_{a^+}^{\alpha}\underline{x}(t, r)$ and ${}^{C}\mathscr{D}_{a^+}^{\alpha}\overline{x}(t, r)$ defined in Definition 4.

Lemma 2 *If $x(t) = (z_1(t), z_2(t), z_3(t))$ is a triangular fuzzy number valued function, then:*

(i) If x is $[(i) - gH]$-differentiable at $t \in [a, b]$ then

$$ ({}_{gH}^{C}\mathscr{D}_{a^+}^{\alpha}x)(t) = ({}^{C}\mathscr{D}_{a^+}^{\alpha}z_1(t), {}^{C}\mathscr{D}_{a^+}^{\alpha}z_2(t), {}^{C}\mathscr{D}_{a^+}^{\alpha}z_3(t)). $$

(ii) If x is $[(ii) - gH]$-differentiable at $t \in [a, b]$ then

$$ ({}_{gH}^{C}\mathscr{D}_{a^+}^{\alpha}x)(t) = ({}^{C}\mathscr{D}_{a^+}^{\alpha}z_3(t), {}^{C}\mathscr{D}_{a^+}^{\alpha}z_2(t), {}^{C}\mathscr{D}_{a^+}^{\alpha}z_1(t)). $$

Definition 7 Let $x : [a, b] \longrightarrow E$ be $[gH]_{\alpha}^{C}$-differentiable at $t \in (a, b)$. We say x is $[(i) - gH]_{\alpha}^{C}$-differentiable at $t \in [a, b]$ if

$$ (i) \quad [({}_{gH}^{C}\mathscr{D}_{a^+}^{\alpha}x)(t)]^r = [{}^{C}\mathscr{D}_{a^+}^{\alpha}\underline{x}(t, r), {}^{C}\mathscr{D}_{a^+}^{\alpha}\overline{x}(t, r)], \ 0 \leq r \leq 1 \tag{5} $$

and that x is $[(ii) - gH]_{\alpha}^{C}$-differentiable at t if

$$ (ii) \quad [({}_{gH}^{C}\mathscr{D}_{a^+}^{\alpha}x)(t)]^r = [{}^{C}\mathscr{D}_{a^+}^{\alpha}\overline{x}(t, r), {}^{C}\mathscr{D}_{a^+}^{\alpha}\underline{x}(t, r)], \ 0 \leq r \leq 1 \tag{6} $$

where

$$ {}^{C}\mathscr{D}_{a^+}^{\alpha}\underline{x}(t, r) = \frac{1}{\Gamma(1-\alpha)}\int_a^t (t-s)^{-\alpha}\frac{d}{ds}\underline{x}(s, r)ds, $$

$$ {}^{C}\mathscr{D}_{a^+}^{\alpha}\overline{x}(t, r) = \frac{1}{\Gamma(1-\alpha)}\int_a^t (t-s)^{-\alpha}\frac{d}{ds}\overline{x}(s, r)ds. $$

Definition 8 ([7]) Let $x : [a, b] \longrightarrow E$ be a fuzzy function. A point $t \in (a, b)$ is said to be a switching point for the $[gH]_\alpha^C$-differentiable of x, if in any neighborhood V of t there exist points $t_1 < t < t_2$ such that:

type I at t_1 (5) holds while (6) does not hold and at t_2 (6) holds and (5) does not holds, or

type II at t_1 (6) holds while (5) does not hold and at t_2 (5) holds and (6) does not holds.

Theorem 3 ([7]) *Let $x : [a, b] \longrightarrow E$ be a fuzzy-valued function on $[a, b]$.*

(i) If x is $[(i) - gH]$-differentiable at $t \in [a, b]$, then

$$[({}_{gH}^{C}\mathscr{D}_{a^+}^{\alpha}x)(t)]^r = [{}^{C}\mathscr{D}_{a^+}^{\alpha}\underline{x}(t, r), {}^{C}\mathscr{D}_{a^+}^{\alpha}\overline{x}(t, r)].$$

(ii) If x is $[(ii) - gH]$-differentiable at $t \in [a, b]$, then

$$[({}_{gH}^{C}\mathscr{D}_{a^+}^{\alpha}x)(t)]^r = [{}^{C}\mathscr{D}_{a^+}^{\alpha}\overline{x}(t, r), {}^{C}\mathscr{D}_{a^+}^{\alpha}\underline{x}(t, r)].$$

3 Main Results

For $\sigma > 0$, we denote by C_σ the space $C([-\sigma, 0], E)$ equipped with the metric defined by

$$D_\sigma[u, v] = \sup_{t\in[-\sigma,0]} D_0[u(t), v(t)].$$

Define $I = [a, a + p]$, $J = [a - \sigma, a] \cup I = [a - \sigma, a + p]$, $p > 0$. Then, for each $t \in I$ we denote by x_t the element of C_σ defined by $x_t(s) = x(t + s)$, $s \in [-\sigma, 0]$.

Consider the following fuzzy fractional functional integral differential equations FFFIDE of order $\alpha \in (0, 1)$ with generalized Hukuhara derivative under form

$$\begin{cases} {}_{gH}^{C}\mathscr{D}_{a^+}^{\alpha}x(t) = f(t, x_t) + \int_a^t g(t, s, x_s)ds, \; t \geq a, \\ x(t) = \varphi(t - a), \; t \in [a - \sigma, a]. \end{cases} \tag{7}$$

where ${}_{gH}^{C}\mathscr{D}_{a^+}^{\alpha}$ is the Caputo's generalized Hukuhara derivative from Definition 6, $f : I \times C_\sigma \longrightarrow E$, $g : I \times I \times C_\sigma \longrightarrow E$ and $\varphi \in C_\sigma$. In this paper, we consider only $[(i) - gH]_\alpha^C$-differentiable type and $[(ii) - gH]_\alpha^C$-differentiable type solutions, i.e, such that there are no switching points in I. By a solution to the initial value problem (7) we mean a fuzzy mapping $x \in C(J, E)$ that satisfies: $x(t) = \varphi(t - a)$ for $t \in [a - \sigma, a]$, x is differentiable on I and ${}_{gH}^{C}\mathscr{D}_{a^+}^{\alpha}x(t) = f(t, x_t) + \int_a^t g(t, s, x_s)ds$, $t \geq a$.

Lemma 3 *Let $\alpha \in (0,1)$, the FFFIDE (7) is equivalent to the following integral equation:*

$$\begin{cases} x(t) = \varphi(t-a),\ t \in [a-\sigma, a], \\ x(t) \ominus_{gH} \varphi(0) = \frac{1}{\Gamma(\alpha)} \int_a^t (t-s)^{\alpha-1} \left(f(s, x_s) + \int_a^s g(s,\tau,x_\tau) d\tau \right) ds,\ t \in [a, a+p]. \end{cases} \tag{8}$$

Now, we consider x and y to be the solutions of Eq. (7) in $[(i)-gH]_\alpha^C$-differentiability type and $[(ii)-gH]_\alpha^C$-differentiability type, respectively, then by using Lemma 3 we have:

$$\begin{cases} x(t) = \varphi(t-a),\ t \in [a-\sigma, a], \\ x(t) = \varphi(0) + \frac{1}{\Gamma(\alpha)} \int_a^t (t-s)^{\alpha-1} \left(f(s, x_s) + \int_a^s g(s,\tau,x_\tau) d\tau \right) ds,\ t \in [a, a+p]. \end{cases} \tag{9}$$

And

$$\begin{cases} y(t) = \varphi(t-a),\ t \in [a-\sigma, a], \\ y(t) = \varphi(0) \ominus \frac{(-1)}{\Gamma(\alpha)} \int_a^t (t-s)^{\alpha-1} \left(f(s, y_s) + \int_a^s g(s,\tau,y_\tau) d\tau \right) ds,\ t \in [a, a+p]. \end{cases} \tag{10}$$

Definition 9 Let $x : J \longrightarrow E$ be a fuzzy function which is $[(i)-gH]_\alpha^C$-differentiable ($[(ii)-gH]_\alpha^C$-differentiable). If x and its derivative satisfy problem (7), we say that x is a (i)-solution ((ii)-solution) of problem (7).

Next, we present main results in this section showing the local existence of two solution for FFFIDE (7). These two solutions correspond, respectively, to $[(i)-gH]_\alpha^C$-differentiability and $[(ii)-gH]_\alpha^C$-differentiability recalled in Definition 7. Contraction mapping theorem is applied to derive local existence and uniqueness of solution.

In the following, for a given constant $k > 0$, we consider the set $X_\varphi^{(i)}$ of all functions $x \in C([a-\sigma, a+p], E)$ such that $x(t) = \varphi(t-a)$ on $[a-\sigma, a]$ and

$$\sup_{t \in [a-\sigma, a+p]} D_0[x(t), \widehat{0}] exp(-k(t+\sigma)) < \infty.$$

In $X_\varphi^{(i)}$, we can define the following metric:

$$D_k^*[x, z] = \sup_{t \in [a-\sigma, a+p]} \{ D_0[x(t), z(t)] exp(-k(t+\sigma)) \},\quad x, z \in X_\varphi^{(i)}.$$

By denoting $\mathbf{S}_k^{(i)} = (X_\varphi^{(i)}, D_k^*)$ we see that $\mathbf{S}_k^{(i)}$ is a complete metric space for $k > 0$.

Next, we consider the FFFIDE (7) under the following assumptions:

$(A1)$ $f \in C([a, a+p] \times C_\sigma, E)$, $g \in C([a, a+p] \times [a, a+p] \times C_\sigma, E)$ and there exists a constant $L > 0$ such that

$$\max\{D_0[f(t,\xi), f(t,\psi)]; D_0[g(t,s,\xi), g(t,s,\psi)]\} \le L D_\sigma[\xi, \psi],$$

for all $\xi, \psi \in C_\sigma$ and $t, s \in [a, a+p]$.

$(A2)$ there exists a constant $M > 0$ such that

$$\max\left\{D_0[f(t,\xi), \widehat{0}]; D_0[g(t,s,\xi), \widehat{0}]\right\} \le M.$$

Theorem 4 *Let the assumptions* $(A1), (A2)$ *holds and the constant:*

$$\delta = \frac{L}{k^\alpha} + \frac{L}{k^{\alpha+1}} - \frac{Lp^\alpha}{k.\Gamma(\alpha+1)} < 1,$$

then, the FFFIDE (7) *has unique solution* x^*.

Proof To proof this theorem we investigate the conditions of Banach fixed point principle.

Define the operator $\mathbf{T} : X_\varphi^{(i)} \longrightarrow X_\varphi^{(i)}$ by:

$$(\mathbf{T}x)(t) = \begin{cases} \varphi(0) + \frac{1}{\Gamma(\alpha)} \int_a^t (t-s)^{\alpha-1} \left(f(s, x_s) + \int_a^s g(s, \tau, x_\tau) d\tau\right) ds, & t \in [a, a+p], \\ \varphi(t-a), \ t \in [a-\sigma, a], \end{cases}$$

Step 1: We shall prove that $\mathbf{T}(X_\varphi^{(i)}) \subseteq X_\varphi^{(i)}$ with assumption $k > b$. Indeed, let $x \in X_\varphi^{(i)}$. For each $t \ge a$, we have:

$$D_0[(\mathbf{T}x)(t), \widehat{0}] = D_0\left[\varphi(0) + \frac{1}{\Gamma(\alpha)} \int_a^t (t-s)^{\alpha-1} \left(f(s, x_s) + \int_a^s g(s, \tau, x_\tau) d\tau\right) ds, \widehat{0}\right],$$

$$\le D_0[\varphi(0), \widehat{0}] + \frac{1}{\Gamma(\alpha)} \int_a^t (t-s)^{\alpha-1} \left(D_0[f(s, x_s), f(s, \widehat{0})] + D_0[f(s, \widehat{0}), \widehat{0}]\right) ds$$

$$+ \frac{1}{\Gamma(\alpha)} \int_a^t (t-s)^{\alpha-1} \left(\int_a^s (D_0[g(s, \tau, x_\tau), g(s, \tau, \widehat{0})] + D_0[g(s, \tau, \widehat{0}), \widehat{0}]) d\tau\right) ds,$$

$$\le D_0[\varphi(0), \widehat{0}] + \frac{1}{\Gamma(\alpha)} \int_a^t (t-s)^{\alpha-1} \left(L D_\sigma[x_s, \widehat{0}] + M\right) ds + \frac{1}{\Gamma(\alpha)} \int_a^t (t-s)^{\alpha-1}$$

$$\left(\int_a^s L D_\sigma[x_\tau, \widehat{0}] d\tau + M(s-a)\right) ds,$$

Further, since $x \in\in X_{\varphi}^{(i)}$, we deduce that there exists $\rho > 0$ such that $D_0[x(t), \widehat{0}] \leq \rho e^{k(t+\sigma)}$, for all $t \in [a-\sigma, a+p]$. Hence $\sup_{\theta\in[-\sigma,0]} D_0[x(t+\theta), \widehat{0}] \leq \rho e^{k(t+\sigma)}$, for all $t \geq a$. In consequence,

$$\sup_{t\in[a,a+p]} \left\{D_0[(\mathbf{T}x)(t), \widehat{0}]e^{-k(t+\sigma)}\right\} \leq$$

$$\sup_{t\in[a,a+p]} \left\{\left(D_0[\varphi(0), \widehat{0}]e^{-k(t+\sigma)} + \frac{\rho L}{\Gamma(\alpha)} \int_a^t (t-s)^{\alpha-1} e^{k(s-t)} ds\right)\right\} + \frac{p^{\alpha} M}{\Gamma(\alpha+1)}$$

$$+ \sup_{t\in[a,a+p]} \left\{\frac{\rho L}{k\Gamma(\alpha)} \int_a^t (t-s)^{\alpha-1} \left(e^{k(s-t)} - e^{k(a-t)}\right) ds\right\} + \frac{Mp^{\alpha+1}}{\Gamma(\alpha+2)},$$

$$\leq \sup_{t\in[a,a+p]} \left\{\left(D_0[\varphi(0), \widehat{0}]e^{-k(t+\sigma)} + \frac{\rho L}{k^{\alpha}} \int_0^{k(t-a)} \frac{u^{\alpha-1}e^{-u}}{\Gamma(\alpha)} du\right)\right\} + \frac{p^{\alpha} M}{\Gamma(\alpha+1)}$$

$$+ \sup_{t\in[a,a+p]} \left\{\frac{\rho L}{k^{\alpha+1}} \int_0^{k(t-a)} \frac{u^{\alpha-1}e^{-u}}{\Gamma(\alpha)} du - \frac{\rho L(t-a)^{\alpha}}{k\Gamma(\alpha+1)}\right\} + \frac{Mp^{\alpha+1}}{\Gamma(\alpha+2)}.$$

Let $H = \sup_{\theta\in[a-\sigma,a]} D_0[\varphi(\theta-a), \widehat{0}]$, then:

$$D_k^*[(\mathbf{T}x)(t), \widehat{0}] \leq H + \frac{\rho L}{k^{\alpha}} + \frac{p^{\alpha} M}{\Gamma(\alpha+1)} + \frac{\rho L}{k^{\alpha+1}} + \frac{p^{\alpha+1} M}{\Gamma(\alpha+2)} - \frac{\rho L p^{\alpha}}{k\Gamma(\alpha+1)}.$$

Besides, for $t \in [a-\sigma, a]$, we get:

$$D_0[(\mathbf{T}x)(t), \widehat{0}]e^{-k(t+\sigma)} = D_0[\varphi(0), \widehat{0}]e^{-k(t+\sigma)} \leq He^{-k(t+\sigma)} \leq He^{-ka}.$$

So that:

$$\sup_{t\in[a-\sigma,a+p]} \left\{D_0[(\mathbf{T}x)(t), \widehat{0}]e^{-k(t+\sigma)}\right\} \leq$$

$$\max\{He^{-ka}, H + \frac{\rho L}{k^{\alpha}} + \frac{p^{\alpha} M}{\Gamma(\alpha+1)} + \frac{\rho L}{k^{\alpha+1}} + \frac{p^{\alpha+1} M}{\Gamma(\alpha+2)} - \frac{\rho L p^{\alpha}}{k\Gamma(\alpha+1)}\} < \infty,$$

and so $(\mathbf{T}x)(t) \in X_{\varphi}^{(i)}$. Hence $\mathbf{T}(X_{\varphi}^{(i)}) \subseteq X_{\varphi}^{(i)}$.

Step 2: The following step, we shall prove that $\mathbf{T}$ is a contraction map by metric D_k^*. For $x, y \in X_{\varphi}^{(i)}$, and $t \in [a-\sigma, a+p]$, we have:

$$D_0[(\mathbf{T}x)(t), (\mathbf{T}y)(t)] = D_0[\varphi(t-a), \varphi(t-a)] = 0, \quad \forall t \in [a-\sigma, a].$$

And

$$D_0[(\mathbf{T}x)(t), (\mathbf{T}y)(t)] = \frac{1}{\Gamma(\alpha)} D_0\left[\int_a^t (t-s)^{\alpha-1} f(s, x_s)ds, \int_a^t (t-s)^{\alpha-1} f(s, y_s)ds\right]$$

$$+ \frac{1}{\Gamma(\alpha)} \int_a^t (t-s)^{\alpha-1} \left(D_0\left[\int_a^s g(s, \tau, x_\tau)d\tau, \int_a^s g(s, \tau, y_\tau)d\tau\right]\right) ds,$$

$$\leq \frac{L}{\Gamma(\alpha)} \int_a^t (t-s)^{\alpha-1} \sup_{\eta\in[-\sigma,0]} D_0[x_s(\eta), y_s(\eta)]ds$$

$$+ \frac{L}{\Gamma(\alpha)} \int_a^t (t-s)^{\alpha-1} \left(\int_a^s \sup_{\eta\in[-\sigma,0]} D_0[x_\tau(\eta), y_\tau(\eta)]d\tau\right) ds,$$

$$= \frac{L}{\Gamma(\alpha)} \int_a^t (t-s)^{\alpha-1} \sup_{\theta\in[s-\sigma,s]} D_0[x(\theta), y(\theta)]ds$$

$$+ \frac{L}{\Gamma(\alpha)} \int_a^t (t-s)^{\alpha-1} \left(\int_a^s \sup_{\theta\in[\tau-\sigma,\tau]} D_0[x(\theta), y(\theta)]d\tau\right) ds.$$

From the definition of the metric D_k^*, it is clear that $D_0[x(s), y(s)] \leq D_k^*[x, y]e^{k(s+\sigma)}$, for all $s \geq a$. Further, for each $t \geq a$, we obtain:

$$D_0[(\mathbf{T}x)(t), (\mathbf{T}y)(t)] \leq$$

$$\frac{L}{\Gamma(\alpha)} \int_a^t (t-s)^{\alpha-1} D_k^*[x, y]e^{k(s+\sigma)}ds + \frac{L}{\Gamma(\alpha)} \int_a^t (t-s)^{\alpha-1} \left(\int_a^s D_k^*[x, y]e^{k(\tau+\sigma)}d\tau\right) ds,$$

$$\leq \frac{LD_k^*[x, y]}{\Gamma(\alpha)} \int_a^t (t-s)^{\alpha-1} e^{k(s+\sigma)}ds + \frac{LD_k^*[x, y]}{k\Gamma(\alpha)} \int_a^t (t-s)^{\alpha-1} e^{k(s+\sigma)}ds - \frac{LD_k^*[x, y]}{k\Gamma(\alpha+1)} e^{k(a+\sigma)},$$

and so:

$$D_K^*[\mathbf{T}x, \mathbf{T}y] = \sup_{t\in[a-\sigma,a+p]} \{D_0[(\mathbf{T}x)(t), (\mathbf{T}y)(t)]exp(-k(t+\sigma))\}$$

$$\leq \frac{LD_k^*[x, y]}{\Gamma(\alpha)} \sup_{t\in[a-\sigma,a+p]} \left\{\int_a^t (t-s)^{\alpha-1} e^{k(s-t)}ds\right\}$$

$$+\frac{LD_k^*[x, y]}{k\Gamma(\alpha)} \sup_{t\in[a-\sigma,a+p]} \left\{\int_a^t (t-s)^{\alpha-1} e^{k(s-t)}ds\right\} - \sup_{t\in[a-\sigma,a+p]} \left\{\frac{LD_k^*[x, y]}{k\Gamma(\alpha+1)} e^{k(a-t)}\right\},$$

$$\leq \frac{L}{k^{\alpha}} D_k^*[x, y] + \frac{L}{k^{\alpha+1}} D_k^*[x, y] - \frac{Lp^{\alpha}}{k\Gamma(\alpha+1)} D_k^*[x, y],$$

$$\leq \left(\frac{L}{k^{\alpha}} + \frac{L}{k^{\alpha+1}} - \frac{Lp^{\alpha}}{k\Gamma(\alpha+1)} \right) D_k^*[x, y].$$

Hence:

$$D_K^*[\mathbf{T}x, \mathbf{T}y] \leq \delta D_k^*[x, y].$$

Since $\delta < 1$, which implies that $\mathbf{T}$ is a contraction map. Consequently, the Banach fixed point principle implies that the FFFIDE (7) has a unique solution x^*.

4 Illustrations

In this section, we shall present some examples being simple illustration of the theory of FFFIDE. We will consider the FFFIDE (7) with $[(i) - gH]_{\alpha}^{C}$-differentiable and $[(ii) - gH]_{\alpha}^{C}$-differentiable, respectively. For this purpose, well consider, for simplicity and without lose of generality, we consider the following FFFIDE:

$$\begin{cases} {}_{gH}^{C}\mathscr{D}_{a^+}^{\alpha} x(t) = f(t, x_t) + \int_a^t k(t, s) x_s ds, \ t \geq a, \\ x(t) = \varphi(t-a), \ t \in [-\sigma, a]. \end{cases} \tag{11}$$

Where $f : I \times C_{\sigma} \longrightarrow E, k : I \times I \longrightarrow \mathbb{R}, \alpha \in (0, 1)$ is the order of the differential equation. Let us denote the r-levels ($r \in [0, 1]$) of x and φ as

$$[x(t)]^r = [\underline{x}(t, r), \overline{x}(t, r)], \quad [\varphi(t)]^r = [\underline{\varphi}(t, r), \overline{\varphi}(t, r)],$$

respectively. Obviously $\underline{x}(., t), \overline{x}(., t) : I \longrightarrow \mathbb{R}$. By using Zadeh's extension principle, we obtain $[f(t, x_t)]^r = [\underline{f}(t, r, \underline{x}(t, r), \overline{x}(t, r)), \overline{f}(t, r, \underline{x}(t, r), \overline{x}(t, r))]$, for $r \in [0, 1]$. In this Eq. (11) we shall solve it by two type of Caputo fractional generalized Hukuhara derivative. Consequently, based on the types of differentiability, we have the following two cases.

Case 1. If $x(t)$ is $[(i) - gH]_{\alpha}^{C}$-differentiable then

$$[({}_{gH}^{C}\mathscr{D}_{a^+}^{\alpha} x)(t)]^r = [{}^{C}D_{a^+}^{\alpha} \underline{x}(t, r), {}^{C} D_{a^+}^{\alpha} \overline{x}(t, r)]$$

and (11) is translated into the following fractional differential system:

$$\begin{cases} {}^{C}D_{a^+}^{\alpha} \underline{x}(t, r) = \underline{\underline{f}}(t, r, \underline{x}(t, r), \overline{x}(t, r)) + \int_a^t \underline{\underline{k(t, s) x_s(r)}} ds, \ t \geq a \\ {}^{C}D_{a^+}^{\alpha} \overline{x}(t, r) = \overline{\overline{f}}(t, r, \underline{x}(t, r), \overline{x}(t, r)) + \int_a^t \overline{\overline{k(t, s) x_s(r)}} ds, \ t \geq a \\ \underline{x}(t, r) = \underline{\varphi}(t-a, r), \ \ \overline{x}(t, r) = \overline{\varphi}(t-a, r) \ t \in [-\sigma, a] \end{cases} \tag{12}$$

Case 2. If $x(t)$ is $[(ii) - gH]_{\alpha}^{C}$-differentiable then

$$[({}_{gH}^{C}\mathscr{D}_{a^+}^{\alpha}x)(t)]^r = [{}^C D_{a^+}^{\alpha}\overline{x}(t,r), {}^C D_{a^+}^{\alpha}\underline{x}(t,r)]$$

and (11) is translated into the following fractional differential system:

$$\begin{cases} {}^C D_{a^+}^{\alpha}\overline{x}(t,r) = \underline{\underline{f}}(t,r,\underline{x}(t,r),\overline{x}(t,r)) + \int_a^t \underline{\underline{k(t,s)x_s(r)}}ds, \ t \geq a \\ {}^C D_{a^+}^{\alpha}\underline{x}(t,r) = \overline{\overline{f}}(t,r,\underline{x}(t,r),\overline{x}(t,r)) + \int_a^t \overline{\overline{k(t,s)x_s(r)}}ds, \ t \geq a \\ \underline{x}(t,r) = \underline{\varphi}(t-a,r), \ \ \overline{x}(t,r) = \overline{\varphi}(t-a,r) \ t \in [-\sigma, a] \end{cases} \tag{13}$$

Where

$$\underline{k(t,s)x_s(r)} = \begin{cases} k(t,s)\underline{\underline{x_s(r)}}, \ if \ k(t,s) \geq 0. \\ k(t,s)\overline{\overline{x_s(r)}}, \ if \ k(t,s) < 0. \end{cases}$$

$$\overline{k(t,s)x_s(r)} = \begin{cases} k(t,s)\overline{x_s(r)}, \ if \ k(t,s) \geq 0. \\ k(t,s)\underline{x_s(r)}, \ if \ k(t,s) < 0. \end{cases}$$

Remark 1 If we ensure that the solution $(\underline{x}(t,r), \overline{x}(t,r))$ of the systems (12) and (13) respectively are valid level sets of fuzzy number valued functions and if the derivatives $({}^C D_{a^+}^{\alpha}\underline{x}(t,r), {}^C D_{a^+}^{\alpha}\overline{x}(t,r))$ are valid level sets of fuzzy number valued functions with two kinds differentiability respectively, then we can construct the solution of the FFFIDE (11).

Example 1 Let us consider the FFFIDE under two kinds of Caputo fractional generalized Hukuhara derivative

$$\begin{cases} ({}_{gH}^{C}\mathscr{D}_{a^+}^{\alpha}x)(t) = x(t-\frac{1}{2}) + \lambda \int_0^t e^{(s-t)}x(s-\frac{1}{2})ds, \ t \in [0, \frac{1}{2}], \\ x(t) = \varphi(t), \ t \in [-\frac{1}{2}, 0]. \end{cases} \tag{14}$$

Where $k(t,s) = \lambda e^{(s-t)}$, $\varphi = (1-t, 2-t, 3-t)$ and $\lambda \in \mathbb{R} - \{0\}$. In this example we shall solve (14) on $[0, \frac{1}{2}]$.

Case 1: $(\lambda > 0 \ \ or \ \ k(t,s) > 0)$

From (4.2), we get

$$\begin{cases} {}^C D_{0^+}^{\alpha}\underline{x}(t,r) = \underline{x}(t-\frac{1}{2},r) + \lambda \int_0^t e^{(s-t)}\underline{x}(s-\frac{1}{2},r)ds, \ t \in [0, \frac{1}{2}], \\ {}^C D_{0^+}^{\alpha}\overline{x}(t,r) = \overline{x}(t-\frac{1}{2},r) + \lambda \int_0^t e^{(s-t)}\overline{x}(s-\frac{1}{2},r)ds, \ t \in [0, \frac{1}{2}], \\ \underline{x}(t,r) = 1+r-t, \ \ \overline{x}(t,r) = 3-r-t \ t \in [-\frac{1}{2}, 0]. \end{cases} \tag{15}$$

By solving (15), we obtain exact solution as follows:

$[x(t)]^r = [1 + r + (1+r)t - \frac{t^2}{2} - \lambda e^{-t}(2+r) + \lambda(2+r-t), 3 - r + (3-r)t - \frac{t^2}{2} - \lambda e^{-t}(4-r) + \lambda(4-r-t)]$.

The solution of (4.4) on $[-\frac{1}{2}, \frac{1}{2}]$ are illustrated in Fig. 1.

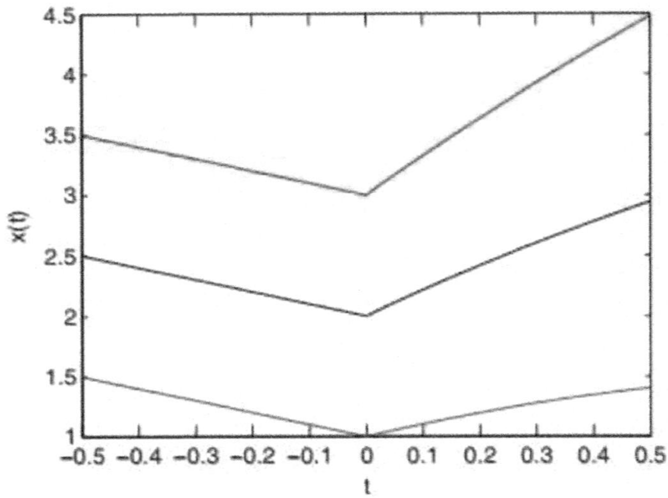

Fig. 1 Graphe of $x(t)$ for $t \in [-\frac{1}{2}, \frac{1}{2}]$, $\lambda = 0.1$

From (13), we obtain:

$$\begin{cases} {}^C D^{\alpha}_{0+}\overline{x}(t,r) = \underline{x}(t-\frac{1}{2}, r) + \lambda \int_0^t e^{(s-t)}\underline{x}(s-\frac{1}{2}, r)ds, \ t \in [0, \frac{1}{2}], \\ {}^C D^{\alpha}_{0+}\underline{x}(t,r) = \overline{x}(t-\frac{1}{2}, r) + \lambda \int_0^t e^{(s-t)}\overline{x}(s-\frac{1}{2}, r)ds, \ t \in [0, \frac{1}{2}], \\ \underline{x}(t,r) = 1 + r - t, \quad \overline{x}(t,r) = 3 - r - t \ t \in [-\frac{1}{2}, 0]. \end{cases} \tag{16}$$

By solving (16), we obtain exact solution as follows:

$[x(t)]^r = [1 + r + (3-r)t - \frac{t^2}{2} - \lambda e^{-t}(4-r) + \lambda(4-r-t), 3 - r + (1+r)t - \frac{t^2}{2} - \lambda e^{-t}(2+r) + \lambda(2+r-t)]$.

The solution of (14) on $[-\frac{1}{2}, \frac{1}{2}]$ are illustrated in Fig. 2.

Case 2: ($\lambda < 0$ *or* $k(t,s) < 0$)

From (12), we get

$$\begin{cases} {}^C D^{\alpha}_{0+}\underline{x}(t,r) = \underline{x}(t-\frac{1}{2}, r) + \lambda \int_0^t e^{(s-t)}\overline{x}(s-\frac{1}{2}, r)ds, \ t \in [0, \frac{1}{2}], \\ {}^C D^{\alpha}_{0+}\overline{x}(t,r) = \overline{x}(t-\frac{1}{2}, r) + \lambda \int_0^t e^{(s-t)}\underline{x}(s-\frac{1}{2}, r)ds, \ t \in [0, \frac{1}{2}], \\ \underline{x}(t,r) = 1 + r - t, \quad \overline{x}(t,r) = 3 - r - t \ t \in [-\frac{1}{2}, 0]. \end{cases} \tag{17}$$

By solving (17), we obtain exact solution as follows:

$[x(t)]^r = [1 + r + (1+r)t - \frac{t^2}{2} - \lambda e^{-t}(4-r) + \lambda(4-r-t), 3 - r + (3-r)t - \frac{t^2}{2} - \lambda e^{-t}(2+r) + \lambda(2+r-t)]$.

The solution of (14) on $[-\frac{1}{2}, \frac{1}{2}]$ are illustrated in Fig. 3.

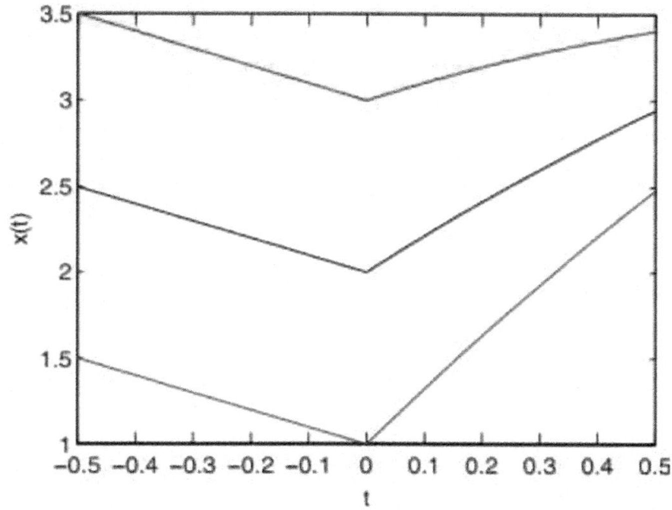

Fig. 2 Graphe of $x(t)$ for $t \in [-\frac{1}{2}, \frac{1}{2}]$, $\lambda = 0.1$

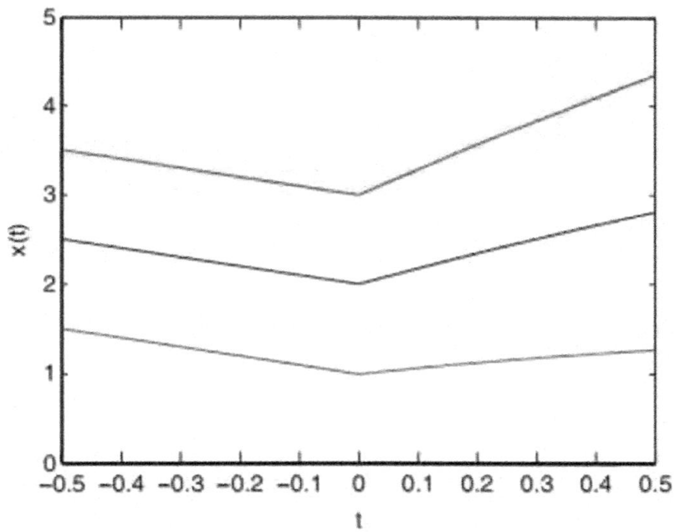

Fig. 3 Graphe of $x(t)$ for $t \in [-\frac{1}{2}, \frac{1}{2}]$, $\lambda = 0.1$

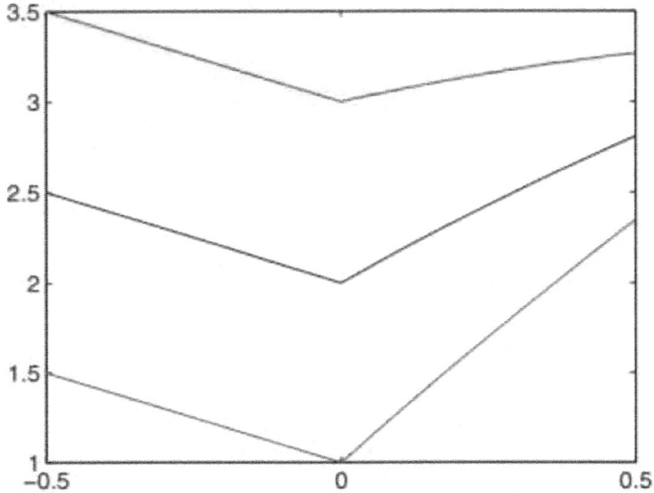

Fig. 4 Graphe of $x(t)$ for $t \in [-\frac{1}{2}, \frac{1}{2}]$, $\lambda = 0.1$

From (13), we get

$$\begin{cases} {}^C D^{\alpha}_{0+} \underline{x}(t, r) = \underline{x}(t - \frac{1}{2}, r) + \lambda \int_0^t e^{(s-t)} \underline{x}(s - \frac{1}{2}, r) ds, \ t \in [0, \frac{1}{2}], \\ {}^C D^{\alpha}_{0+} \overline{x}(t, r) = \overline{x}(t - \frac{1}{2}, r) + \lambda \int_0^t e^{(s-t)} \overline{x}(s - \frac{1}{2}, r) ds, \ t \in [0, \frac{1}{2}], \\ \underline{x}(t, r) = 1 + r - t, \quad \overline{x}(t, r) = 3 - r - t \ t \in [-\frac{1}{2}, 0]. \end{cases} \tag{18}$$

By solving (18), we obtain exact solution as follows:

$[x(t)]^r = [1 + r + (3 - r)t - \frac{t^2}{2} - \lambda e^{-t}(2 + r) + \lambda(2 + r - t), \quad 3 - r + (1 + r)t - \frac{t^2}{2} - \lambda e^{-t}(4 - r) + \lambda(4 - r - t)]$.

The solution of (14) on $[-\frac{1}{2}, \frac{1}{2}]$ are illustrated in Fig. 4.

5 Conclusions

In this paper, we have obtained a local existence and uniqueness result for a solution to fuzzy fractional functional integration and differential equations using the concept of Caputo generalized Hukuhara differentiability. In this setting, we prove the existence of two fuzzy solutions, each one corresponding to a different type of derivative but without using switching points by the contraction principle.

References

1. Agarwal, R.P., Lakshmikantham, V., Nieto, J.J.: On the concept of solution for fractional differential equations with uncertainty. Nonlinear Anal. (TMA) **72**, 2859–2862 (2010)
2. Agarwal, R.P., Arshad, S., O'Regan, D., Lupulescu, V.: Fuzzy fractional integral equations under compactness type condition. Fract. Calcul. Appl. Anal. **15**, 572–590 (2012)
3. Ahmad, M.Z., Hasan, M.K.: A new approach to incorporate uncertainty into Euler's method. Appl. Math. Sci. **4**, 2509–2520 (2010)
4. Ahmad, M.Z., Hasan, M.K., Baets, B.D.: Analytical and numerical solutions of fuzzy differential equations. Inf. Sci. **236**, 156–167 (2013)
5. Ahmad, M.Z., Hasan, M.K., Abbasbandy, S.: Solving fuzzy fractional differential equations using Zadeh's extension principle. **2013**, Article ID 454969, 11 (2013). http://dx.doi.org/10.1155/2013/454969
6. Alikhani, R., Bahrami, F.: Global solutions for nonlinear fuzzy fractional integral and integrodifferential equations. Commun. Nonlinear Sci. Numer. Simul. **18**, 2007–2017 (2013)
7. Allahviranloo, T., Gouyandeh, Z., Armand, A.: Fuzzy fractional differential equations under generalized fuzzy Caputo derivative. J. Intell. Fuzzy Syst. **26**, 1481–1490 (2014)
8. Allahviranloo, T., Salahshour, S., Abbasbandy, S.: Explicit solutions of fractional differential equations with uncertainty. Soft Comput. Found Meth. Appl. **16**, 297–302 (2012)
9. Allahviranloo, T., Abbasbandy, S., Sedaghatfar, O., Darabi, P.: A new method for solving fuzzy Integro-differential equation under generalized differentiability. Neural Comput. Appl. **21**, 191–196 (2012)
10. Allahviranloo, T., Kiani, N.A., Motamedi, N.: Solving fuzzy differential equations by differential transformation method. Inf. Sci. **179**, 956–966 (2009)
11. Allahviranloo, T., Abbasbandy, S., Salahshour, S., Hakimzadeh, A.: A new method for solving fuzzy linear differential equations. Computing **92**, 181–197 (2011)
12. Arshad, S., Lupulescu, V.: On the fractional differential equations with uncertainty. Nonlinear Anal. (TMA) **74**, 85–93 (2011)
13. Bede, B., Gal, S.G.: Generalizations of the differentiability of fuzzy-number-valued functions with applications to fuzzy differential equations. Fuzzy Sets Syst. **151**, 581–599 (2005)
14. Bede, B., Rudas, I.J., Bencsik, A.L.: First order linear fuzzy differential equations under generalized differentiability. Inf. Sci. **177**, 1648–1662 (2007)
15. Bede, B.: A note on 'two-point boundary value problems associated with non-linear fuzzy differential equations'. Fuzzy Sets Syst. **157**, 986–989 (2006)
16. Bede, B., Tenali, G.B., Lakshmikantham, V.: Perspectives of Fuzzy Initial Value Problems. Commun. Appl. Anal. **11**, 339–358 (2007)
17. Bede, B., Stefanini, L.: Generalized differentiability of fuzzy-valued functions. Fuzzy Sets Syst. **230**, 119–141 (2013)
18. Diethelm, K.: The Analysis of Fractional Differential Equations (An Application Oriented Exposition Using Differential Operators of Caputo Type). Lecture Notes in Mathematics. Springer, Berlin, Heidelberg (2004)
19. Fard, O.S., Salehi, M.: A survey on fuzzy fractional variational problems. J. Comput. Appl. Math. **271**, 71–82 (2014)
20. Gasilov, N.A., Fatullayev, A.G., Amrahov, S.E., Khastan, A.: A new approach to fuzzy initial value problem. Soft Comput. **18**, 217–225 (2014)
21. Gnana Bhaskar, T., Lakshmikantham, V., Leela, S.: Fractional differential equations with a Krasnoselskii-Krein type condition. Nonlinear Anal.: Hybrid Syst. **3**, 734–737 (2009)
22. Hale, J.K.: Theory of Functional Differential Equations. Springer, New York (1977)
23. Hoa, N.V., Phu, N.D.: Fuzzy functional integro-differential equations under generalized H-differentiability. J. Intell. Fuzzy Syst. **26**, 2073–2085 (2014)
24. Hoa, N.V., Tri, P.V., Dao, T.T.: Some global existence results and stability theorem for fuzzy functional differential equations. J. Intell. Fuzzy Syst. (Inpress)
25. Hukuhara, M.: Integration des applications mesurables dont la valeur est un compact convex. Funkc. Ekvacioj **10**, 205–229 (1967)

26. Kaleva, O.: A note on fuzzy differential equations. Nonlinear Anal. **64**, 895–900 (2006)
27. Khastan, A., Nieto, J.J.: A boundary value problem for second order fuzzy differential equations. Nonlinear Anal.: Theory Methods Appl. **72**, 3583–3593 (2010)
28. Khastan, A., Nieto, J.J., Rodriguez-Lopezez, R.: Variation of constant formula for first order fuzzy differential equations. Fuzzy Sets Syst. **177**, 20–33 (2011)
29. Khastan, A., Nieto, J.J., Rodriguez-Lopez, R.: Fuzzy delay differential equations under generalized differentiability. Inform. Sci. **275**, 145–167 (2014)
30. Khastan, A., Nieto, J.J., Rodriguez-Lopez, R.: Schauder fixed-point theorem in semilinear spaces and its application to fractional differential equations with uncertainty. Fixed Point Theory Appl. **2014**, 21 (2014). https://doi.org/10.1186/1687-1812-2014-21
31. Kilbas, A.A., Marzan, S.A.: Cauchy problem for differential equation with Caputo fractional derivative. Fract. Calc. Appl. Anal. **7**, 297–321 (2004)
32. Kilbas, A.A., Srivastava, H.M., Trujillo, J.J.: Theory and applications of fractional differential equations. North-Holland Mathematics Studies, vol. 204. Elsevier, Amsterdam (2006)
33. Kilbas, A.A., Srivastava, H.M., Trujillo, J.J.: Theory and Applications of Fractional Differential Equations. Elsevier Science B.V, Amsterdam (2006)
34. Kuang, Y.: Delay Differential Equations with Applications in Population Dynamics. Academic, Boston (1993)
35. Lakshmikantham, V., Leela, S.: A Krasnoselskii-Krein-type uniqueness result for fractional differential equations. Nonlinear Anal.: TMA **71**, 3421–3424 (2009)
36. Lakshmikantham, V., Mohapatra, R.N.: Theory of Fuzzy Differential Equations and Applications. Taylor and Francis, London (2003)
37. Lakshmikantham, V.: Theory of fractional functional differential equations. Nonlinear Anal.: Theory Methods Appl. **69**, 3337–3343 (2008)
38. Lupulescu, V.: On a class of fuzzy functional differential equations. Fuzzy Sets Syst. **160**, 1547–1562 (2009)
39. Lupulescu, V.: Fractional calculus for interval-valued functions. Fuzzy Set Syst. (accepted) (2013)
40. Lupulescu, V.: On a class of functional differential equations in Banach spaces. Electr. J. Qual. Theory Diff. Eq. (64), 1–17
41. Malinowski, M.T.: Random fuzzy differential equations under generalized Lipschitz condition. Nonlinear Anal.: Real World Appl. **13**, 860–881 (2012)
42. Malinowski, M.T.: Existence theorems for solutions to random fuzzy differential equations. Nonlinear Anal.: Theory Methods Appl. **73**, 1515–1532 (2010)
43. Malinowski, M.T.: Second type Hukuhara differentiable solutions to the delay setvalued differential equations. Appl. Math. Comput. **218**, 9427–9437 (2012)
44. Mazandarani, M., Kamyad, A.V.: Modified fractional Euler method for solving fuzzy fractional initial value problem. Commun. Nonlinear Sci. Numer. Simul. **18**, 12–21 (2013)
45. Mazandarani, M., Najariyan, M.: Type-2 fuzzy fractional derivatives. Commun. Nonlinear Sci. Numer. Simul. **19**, 2354–72 (2014)
46. Nieto, J.J., Khastan, A., Ivaz, K.: Numerical solution of fuzzy differential equations under generalized differentiability. Nonlinear Anal.: Hybrid Syst. **3**, 700–707 (2009)
47. Odibat, Z.M., Shawagfeh, N.T.: Generalized Taylor's formula. Appl. Math. Comput. **186**, 286–293 (2007)
48. Odibat, Z.M.: Approximations of fractional integrals and Caputo fractional derivatives. Appl. Math. Comput. **178**, 527–533 (2006)
49. Podlubny, I.: Fractional Differential Equation. Academic, San Diego (1999)
50. Salahshour, S., Allahviranloo, T., Abbasbandy, S., Baleanu, D.: Existence and uniqueness results for fractional differential equations with uncertainty. Adv. Diff. Eq. **2012**, 112 (2012)
51. Salahshour, S., Allahviranloo, T., Abbasbandy, S.: Solving fuzzy fractional differential equations by fuzzy Laplace transforms. Commun. Nonlinear Sci. Numer. Simul. **17**, 1372–1381 (2012)
52. Tri, P.V., Hoa, N.V., Phu, N.D.: Sheaf fuzzy problems for functional differential equations. Adv. Diff. Eq. **2014**, 156 (2014). https://doi.org/10.1186/1687-1847-2014-156

53. Vu, H., Dong, L.S., Hoa, N.V.: Random fuzzy functional integro-differential equations under generalized Hukuhara differentiability. J. Intell. Fuzzy Syst. **27**, 1491–1506 (2014)
54. Vu, H., Hoa, N.V., Phu, N.D.: The local existence of solutions for random fuzzy integrodifferential equations under generalized H-differentiability. J. Intell. Fuzzy Syst. **26**, 2701–2717 (2014)

Social Dilemmas and the Emergence of Cooperation in Financing Public Goods

Miloudi Kobiyh and Slimane Ed-Dafali

Abstract Social dilemmas in economics characterize situations of social interaction in which the contradiction between individual and collective rationality occur in the classical game theory. In the case of financing public good, analyzed from the viewpoint of the Prisoner's Dilemma to many players, this study highlights how social preferences determine the emergence of cooperation in accordance with the different results of experimental studies. The current study contributes to the literature with the aim of broadening our sight of the emergence of collaborative behaviors by taking into account the question of the nature of individual preferences and seeking to integrate the intentions and emotions of the players when seeking a conditionally efficient equilibrium which could be qualitatively greater than the Pareto optimal allocation. This paper provides a comprehensive analysis and discussion of the importance of adopting a collaborative behavior for providing an efficient solution to the problem of financing public goods when social dilemmas arise.

Keywords Voluntary contribution · Public goods · Prisoner's Dilemma · Collaborative behaviors · Social preferences

1 Introduction

Social dilemmas characterize situations of interactions of economic agents where a conflict can arise between the pursuit of individual interest and the search for collective interest. These situations are often illustrated by games where individual rationality prevents economic agents from cooperating. However, a multitude of experimental works have revealed the presence of cooperative behaviors in these games for which standard theory predicts the existence of opportunistic behaviors (see, [5]). Obviously, the results of these experimental studies violate the traditional

M. Kobiyh · S. Ed-Dafali (✉)
LIRO Laboratory, ENCGJ, University of Chouaïb Doukkali El Jadida, El Jadida, Morocco
e-mail: slimane.eddafali@gmail.com

S. Melliani et al. (eds.), *Applied Mathematics and Modelling in Finance, Marketing and Economics*, Studies in Computational Intelligence 1114,
https://doi.org/10.1007/978-3-031-42847-0_10

assumptions of selfishness and rationality and challenge the traditional view of human behavior (see, [4, 6, 7]).

Indeed, research in experimental economics is developing around various issues, including that of identifying the motivations for cooperation between agents (see, [2, 8]). The Prisoner's Dilemma game illustrates the idea that the confrontation of individual interests does not necessarily lead to a collective optimum. However, the multiple experiments carried out on the Prisoner's Dilemma game showed that there is some cooperation between the players. Furthermore, the experimental results revealed the presence of cooperative behaviors in games for which standard theory predicts the existence of selfish behaviors.

Cooperation is the stake of most economic, political and social relations. For instance, firms have an interest in agreeing on high prices, states have an interest in making peace and developing trade, and employers and employees have an interest in maintaining a relationship of trust. However, seeking private interests is not beneficial for cooperation. Economic analysis often does not retain exchanges between individuals only in the case of purely commercial links. However, individuals do not only exchange market goods or services, but also often intangible qualities that play an important role in the structuring of social and economic activities.

The role of prosocial behaviors is preponderant in explaining the gap between theoretical prediction and experimental observations. In this regard, research on social preferences and their role in different individual interactions has become the natural theoretical background for studies on cooperation games. This article is organized as follows. In the first section, we study the role of social preferences in social dilemmas by presenting theoretical modeling. In the second section, we present the classical theory of public goods. We then integrate social preferences by explaining their role in the emergence of the contribution. Finally, we discuss the role of emotions in social preferences in the third section.

2 Social Dilemmas and Social Preference Models

2.1 *The Role of Social Preferences*

People are not motivated only by money. Research in social psychology has revealed that social and moral dimensions are a strong motivator alongside these material motivations. Social preferences are preferences that are distinct from personal interest. They also explain the search for individual interest. It is said that a person has social preferences if she cares not only about his material gain but also that of others.

The explanation of individual behavior and the exploration of patterns have been experimented with using sophisticated materials (computers or other communication devices). Experimentation and its protocols have become a mode of exploration of economic attitudes and behaviors leading to the development of an experimental economy. This method makes it possible to isolate certain contexts, presented by theory as determinants of these behaviors.

Research in experimental economics is developing around various issues, including that of identifying the motivations for cooperation between agents. However, individual preferences differ from personal interest (see, [2, 8]). It also is important to point out that In recent years, several theories on social preferences have been developed (see, [4, 6, 7, 9, 18]).

The results of such studies violated traditional assumptions of selfishness and rationality and challenge the traditional view of human behavior. Thus, a pure rational decision theory is insufficient to take into consideration economic behaviors since defection will not always be the behavior followed by people as indicated in classical theory.

Due to the presence of positive externalities, the social value of cooperation also increases. Based on these observations, a literature has been developed on the intrinsic motivations of individuals and their spirit of cooperation. (see, [16]). In this regard, cooperation depends on several cultural, psychological and emotional characteristics, in addition to the conditional cooperation (see, [10]), and moral and social motivations such as benevolence, empathy, altruism and fairness. It is within the theoretical models of social preferences that moral emotions can be taken into account in economic theory, these emotions can have behavioral effects (see, [14]). It has thus been shown that emotions, such as anger, regret and guilt, play an important role in decision-making (see, [17]). In order to highlight the way in which emotions and rationality are combined in a reasoned process, a theoretical support for these motivations finds its legitimacy in the theories dealing with intentions which account for fairness and reciprocity (see, [7, 18]). To this end, reciprocity can contribute to the emergence and support of cooperation (see, [17]).

Moreover, the influence of emotions is exerted in different ways depending on the personality of individuals, and mainly according to their degree of social and moral orientation, in addition to the importance of the emotions felt such as guilt or regret. From this perspective, (see, [14]) highlighted the link between empathy and behavior in order to illustrate the relationship between emotion and social orientation. In this sense, empathy is considered a fundamental requirement of any prosocial behavior.

Likewise, (see, [14]) highlighted also the role of social consciousness in the development of prosocial behavior. In this sense, economic agents often have an empathetic personality and tend to adopt prosocial behaviors (see, [14]). The cognitive and affective dimension of empathy can have different functions in the emergence and development of prosocial behaviors, it includes the ability to understand what others are thinking or feeling (see, [1]).

Therefore, emotions are seen as exogenous preferences and are activated because social and moral norms implicitly pre-exist and constrain individual choices. An interpretation in terms of beliefs assumes that people make predictions about the will of their partners. In other words, a player will have beliefs about a player's behavior whether they engage in selfish or prosocial behavior.

Moreoever, emotions considerably influence individual's choices when it comes to social dilemmas (see, [17]). From this perspective, the way of integrating emotions relies on the most visible characteristic of emotions, distinguishing positive emotions from negative ones, and integrating these emotions as exogenous parameters in the traditional utility function (see, [14]).

2.2 *Social Preference Models*

Several researchers have attempted to formally integrate these moral and social motivations into utility theory by proposing models of social preferences. One of the great merits of these models of social preferences lies in their ability to express the conditions under which the existence of individuals with these preferences influences the balance of a set of social interactions.

2.2.1 Rabin's (1993) Fairness Model

Rabin [18] pointed to the role of intentions as the source of fairness and reciprocity behavior. His model captures the fact that individuals are willing to sacrifice a fraction of their payment to help those who have behaved generously towards them. Likewise, individuals are prepared to punish those who have behaved in a hostile manner towards them. Rabin proposed the concept of a fairness equilibrium to account for these facts.

Thus, he has built a game in which the utility function of player i depends not only on his monetary gain but also on intentional fairness. The arguments of this utility are therefore a_i, b_j and c_i such that a_i the strategy of player i, b_j represents the belief of player i on the strategy of player j and c_i the belief of player i on the belief of player j on his own strategy.

The expression of this utility function is as follows:

$$U_i(a_i, b_j, c_i) = \pi_i(a_i, b_j) + h_j(b_j, c_i)[1 + f_i(a_i, b_j)]$$

The variable π_i corresponds to the utility obtained by the material gain of the player, the term $h_j(1 + f_i)$ is the utility part U_i which attempts to capture intentional fairness. Thus, if player i considers that player j will adopt a benevolent behavior ($h_j > 0$), then he will be benevolent towards him ($f_i > 0$).

On the other hand, if player i thinks that player j is going to behave maliciously ($h_j < 0$), then he will also be malicious towards him ($f_i < 0$). It turns out that factoring in this intentional fairness can increase or decrease overall utility. If the sign of the term $h_j(1 + f_i)$ is negative, it means that $U_i \leq \pi_i$.

Therefore, a player's utility comes not only from his monetary gain but also from the intentional fairness of the situation as he sees it. The concept of fairness equilibrium assumes that expectations about the intentions of the players turn out to be correct. It is defined as a set of strategies and beliefs about the intentions of other players such that no player has an interest in unilaterally changing strategies, considering the strategies of other players and their own beliefs about their intentions.

Rabin's model is pioneering in terms of providing the fairness equilibrium and in capturing the mutual intentions. However, Rabin clarifies that a fair balance is not necessarily a Nash equilibrium.

2.2.2 The Fehr and Schmidt's (1999) Model of Inequity Aversion

The Fehr-Schmidt model assumes a population of individuals who do not like to end up with a monetary gain less than the average gain and who no longer like to benefit from a gain greater than the average gain (see, [9]). Consider n players and the vector of gains $g = (g_i, g_{-i})$ is a common knowledge, the utility function of a player i benefiting from the gain g_i may be represented as follows:

$$U_i(g) = g_i - \frac{\alpha_i}{n-1}\sum_{j\neq i}\max\{g_j - g_i, 0\} - \frac{\beta_i}{n-1}\sum_{j\neq i}\max\{g_i - g_j, 0\}$$

With $\beta_i \leq \alpha_i$ and $0 \leq \beta_i < 1$.

In this function, the first term is the gain of player i which measures the maximum level of his utility. The second term measures the dissatisfaction that results from a psychological loss perceived by an individual when his gain is lower than that of others. The coefficient α_i measures the envy which evaluates his suffering when he obtains less than the others. The third term also represents dissatisfaction resulting from a gain greater than that of other players. The coefficient β_i measures the empathy or compassion he feels towards others.

The $\beta_i \leq \alpha_i$ inequality means that envy is stronger than empathy. The condition $0 \leq \beta_i$ means that player i derives no satisfaction from the superiority of his gain. The condition $\beta_i < 1$ rules out the possibility that the loss of utility is quite high in the case of empathy.

This model postulates that individuals are motivated by Fairness and inequity aversion and are willing to sacrifice part of their earnings to reduce inequity. However, these individuals are only concerned with the fairness of their own material payment relative to the payments of others. This causes aversion to inequity to take the form of aversion to inequality.

2.2.3 Bolton and Ockenfels' (2000) Equity, Reciprocity and Competition-ERC Model

The emphasis in this model is on relative player earnings (see, [3]). It incorporates the idea of relative income as individuals make their choices taking into account not only their own earnings but also the earnings of others. This is a situation of choice according to a process of social comparison, since the player attaches utility to his own gain and to his position in relation to other players. The behavior of player i is assumed to be dictated in this model by the maximization of his motivation function ν_i which depends on his gain g_i and the relative gain $\frac{g_i}{\sum_k g_k}$:

$$\nu_i = \nu_i\left(g_i, \frac{g_i}{\sum_k g_k}\right)$$

where $\frac{g_i}{\sum_k g_k}$ represents the relative share of player i in the final payment.

This motivation function represents the utility function of the player whose reference point is equal sharing: the player has a preference for a relative gain equal to the average gain. It reflects what drives the player's behavior. Thus, the player does not only attach utility to his gain but also to his situation in relation to other players. In other words, the player positively perceives an increase in his relative gain when it is below the average. Ultimately, this is a fair behavior since people do not want to be treated less well than others. Likewise, these individuals are willing to sacrifice a fraction of their wealth to help others who are worse off than themselves, reflecting reciprocal behavior.

3 Social Preferences in Public-Good Games

3.1 The Public Goods Game

The public good game is a generalization of the Prisoner's Dilemma, as it highlights the same conflict between the collective interest based on cooperation and the individual interest which does not require cooperation. It is based on a more general framework, involving more than two players and more than two strategies for each player. The main characteristic of a public good is that its consumption by one economic agent does not prevent its consumption by another agent. Moreover, it is impossible to exclude an economic agent from the consumption of this good. In this sense, the provision of this good comes up against the phenomenon of stowaway behavior, since it is indeed rational for a homo conomicus to seek to benefit from a public good without contributing to its funding (see, [16]). In this research, we are interested in the voluntary and individual contribution to the financing of public goods. This shows the importance of social preferences and conditional contribution in supporting this voluntary funding. These preferences revealed the ability of moral, social and emotional elements to support contribution to the public good (see, [16]). Several examples illustrate the voluntary contribution to the financing of a public good such as associations and charities, or the national solidarity days and the highest individual effort in a team.

This game has been studied and tested from the 1980s by several authors (e.g. [11–13, 15, 17]). The study of the decision to contribute to the financing of public goods is closely associated with the understanding of the emergence of cooperation in societies. In fact, in addition to its properties of non-rivalry and non-exclusion, a public good is characterized by a marginal return per capita lower than that of a private good (see, [17]).

This game can be presented by a simplified experimental protocol. Consider a game of four players ($n = 4$) who do not have the possibility of communicating with each other and each of whom has an endowment of ten tokens $D_i = 10$, he can either keep them or contribute with a number of tokens c_i form a common pot (the tokens kept will be invested in a private good).

The gain function of each is defined as follows:

$$g_i = 2(D_i - c_i) + \Sigma_{k=1}^{4} c_k$$

In this gain function, the marginal per capita return of the public good is equal to 1 for simplicity. And the number of tokens kept will be multiplied by 2 which is the marginal return of the private good (obviously the marginal return of the private good is greater than the marginal return on the public good). It is also obvious that this situation represents a social dilemma insofar as the only dominant strategy, which is the Nash equilibrium of this game, corresponds to a zero contribution to the public good: $c_i = 0$. And this because:

$$\frac{dg_i}{dc_i} = -2 + 1 = -1 < 0$$

whereas the social optimum requires a contribution by all the participants and with all the endowment ($c_i = D$). For example, if one player keeps 6 tokens and the other three place 12 tokens in all, then the player's payout is $G = 2.6 + 1.12 = 24$. It is indeed a game of the Prisoner's Dilemma generalized to more than two players and to more than two strategies (each player can contribute with a number of tokens between 0 and 10, so 11 ways to contribute which gives rise to 11 strategies).

The standard behavior is rational egoism which maximizes individual gain: no one contributes to the public good and everyone receives the gain $G = 20$.

Indeed, a token kept earns a gain of 2, while a token placed in the jackpot only pays a gain of 1. Therefore, any strategy of placing tokens in the jackpot will be dominated by the strategy of holding them. Suppose a player contributes when his partners do not, he will have 0 + 1.10 = 10. But if all players place their tokens, each will win 2.0 + 1.40 = 40. It is obvious that this situation cannot correspond to a Nash equilibrium because each player has an interest in unilaterally changing his strategy by lowering his own contribution. Thus, given that the other participants place their ten tokens, placing 10 tokens provides a gain of 40 while keeping saves 2.10 + 1.30 = 50. It turns out that each player has an interest in minimizing their contribution. Thus, each participant keeps the ten tokens, so no one contributes to the public good and each receives the gain $G = 20$.

Therefore, by considering the two situations: either all the participants contribute with the 10 tokens or do not do so, the gains of one player among the four can be summarized in the following Table 1.

Table 1 Comparison of player gains by voluntary contribution decision (0 or 10)

	Contribute (10)	Do not contribute (0)
Contribute (10)	40	10
Do not contribute (0)	50	20

From the Table 1, non-contribution is the dominant strategy. And the Nash equilibrium according to the game's standard theoretical solution will be the zero contribution to the public good. This reflects the behavior of the stowaway highlighted by standard economic theory. Everyone wants to enjoy the good without contributing believing that they are the only ones doing it which results in not reaching the social optimum because everyone will get 20 instead of 40. This means that individuals attempt to profit from the public good without participating in its financing, reflecting a fundamental conflict between the individual interest in not contributing and the social incentive to contribute.

However, experimental observations call into question the theoretical prediction that the behavior of individuals is purely selfish and socially suboptimal. Experimental studies show that the contribution is between 40% and 60% of the initial individual endowment (see, [15]). On the other hand, if the game is repeated, the average contribution is 50% but which slows down over the repetitions (see, [13]).

The existence of social preferences can give rise to contribution to the public good and induce reluctant participants to contribute. Several sets of social interactions conclude in favor of the arguments of reciprocity and aversion to inequity in the contribution to the public good (see, [7]). This shows the role and implication of prosocial behaviors (e.g. altruism, cooperation, trust, fairness).

3.2 Towards Public-Good Games with Social Preferences Theories

We have already seen that social preference models can explain cooperative or non-cooperative behaviors among players. These models highlight the heterogeneity of preferences that interacts significantly with the economic environment and determine the nature of the emotions that are involved and shape behavior.

In the game of public good, a single player with selfish behavior is able to get other players not to contribute. On the other hand, the existence of individuals who are averse to inequity can influence the outcome of the game by causing individuals to contribute. Likewise, by comparing his earnings to that of other players, the individual suffers both from inequalities in his favor (guilt) and inequalities against him (envy), but he suffers more from inequalities playing against him. Thus, in view of their rationality and aversion to inequality, players can punish other individuals to reduce the inequality of earnings resulting from unequal contributions.

Table 2 Comparison of player gains, including the role of emotions

	Contribute (10)	Do not contribute (0)
Contribute (10)	40	$10\theta_i - 30\omega_i$
Do not contribute (0)	$30(\theta_k + \omega_k)$	$20\lambda_i$

Ultimately, the economic environment determines the type of preferences that is decisive in inducing the behavior that prevails at equilibrium. Depending on these preferences and their heterogeneity, the contribution has every chance of being a game-changer. These preferences that capture the intentions and emotions of the players can be modeled by a utility function that expresses the set of elements that determine the behavior of the participants (Table 2).

Thus, a general framework for the formalization of social preferences can be put forward to explain the choice of actors in terms of financing the public good. The emotional consequences associated with each state of nature will need to be factored into the utility calculation. In addition, the introduction of emotion parameters such as satisfaction and regret is necessary to explain the choice of participants. This utility captures, in addition to intentions about the behavior of others, the player's satisfaction with all the contributions and his situation of inferiority or superiority in relation to others. This individual utility function is defined as follows:

$$U(c_i, c_{-i}) = U\left[2(D_i - c_i) + \sum_{k=1}^{4} c_k\right] = 2U(D_i - c_i) + U\left(\sum_{k=1}^{4} c_k\right)$$

We have assumed that the utility function is additive.

Suppose further that $2U(D_i - c_i) = 2\lambda(D_i - c_i)$, in the same way $U(\sum_{k=1}^{4} c_k) = \theta \sum_{k=1}^{4} c_k + \omega \sum_{j \neq i}(c_j - c_i)$, the utility function of a player is:

$$U(c_i, c_{-i}) = 2\lambda(D_i - c_i) + \theta \sum_{k=1}^{4} c_k + \omega \sum_{j \neq i}(c_j - c_i)$$

The parameters λ, θ and ω are emotion parameters which capture respectively the satisfaction of the player of his own choice, his satisfaction with the contribution of all the players, and the regret that could result from the differences between his contribution and the contributions of others. λ and θ are assumed to be positive emotion parameters as long as they capture player's satisfaction, they belong to the interval [0,1]. On the other hand, the parameter ω captures a negative emotion, namely the regret that the player feels when there are gaps between the other contributions and his own, so this parameter belongs to the interval [–1.0].

In the table of gains associated with the two choices (Contribute by the totality of the endowment or do not contribute), we now include the anticipated valence of the emotion in the same way as the gains. This is associated with the feeling as a result of the consequences to the two choices by all players. Thus, a player with emotions that are expressed in his utility function indicates that he is behaving prosocial. So, let's go back to the public good game and the payments matrix in Table 1, and consider a participant with this prosocial behavior and only two options ($c_i = 10 \ or \ 0$).

When $c_i = 10$ and $c_k = 10$ (i.e. everyone contributes with the entire endowment), the utility is $U = 40$. When the player behaves prosocial, he prefers to contribute and he is fully satisfied to do so when others contribute, thus $\theta = 1$ and $U = 40$. If $c_i = 10$ and $c_k = 0$ (he alone who contributed), in this case the utility is $10\theta_i - 30\omega_i$ because the satisfaction which results from all the contributions will be only $\theta_i \neq 0$ and he feels regretful $\omega_i \neq 0$ differences in contributions.

If $c_i = 0$ and $c_k = 0$ which means that no participant contributes, we will only have the satisfaction of the player's choice $\lambda_i \neq 0$: when everyone opts for no contribution, the player's utility is $20\lambda_i$. In the case where $c_i = 0$ and $c_k = 10$ (the others have contributed and he has not), the utility is $20\lambda + 30\theta_k + 30\omega_k$, with $\lambda = 0$ since he has no satisfaction of his choice, therefore the utility will be $30\theta_k + 30\omega_k$: he will have no satisfaction with his choice not to contribute but will have some satisfaction with the contribution of others $\theta_k \neq 0$ and he feels regretful at the deviations of the contributions $\omega_k \neq 0$.

Then we have the following payoff matrix.

When considering the values of emotion parameters, non-contribution cannot always be a dominant strategy. So, if the player adopts a selfish attitude, he prefers not to contribute and he is happy to do so. If this is the behavior of the population, it is obvious that the non-contribution will be the balance of the game. However, the observed contribution reflects that individuals do not exhibit this behavior. Emotions, which are involved in the game situation, influence the decision to contribute according to the emotional intensity felt by the participants. As $(\theta_k + \omega_k) \leq 1$ because the parameter ω_k belongs to the interval [-1,0], utility $30(\theta_k + \omega_k)$ is always less than 40, so non-contribution will never be a dominant strategy. To this end, the contribution can be accepted as an equilibrium of the game but is not a dominant strategy. Likewise, non-contribution can also be an equilibrium of the game. However, in the case where $10\theta_i - 30\omega_i > 20\lambda_i$, the contribution will be the dominant strategy and therefore the only equilibrium of the game. This shows the vital role that emotions play in individual choices and therefore in the balance of the game. Therefore, the contribution can be the equilibrium of the game which explains the observed contribution even in a one-shot game.

This refers to the strong anticipation of contributions from others, which highlights the important and decisive role of intentions and conditional contribution. This conditional contribution highlighted a situation in which the contributions of others influence subject contributions. Consequently, this explains the contributions of players and that emotions are involved and generate non-satisfaction in the absence of contribution. The regret of not contributing will be so high, which may reduce the utility perceived by the players in the event of dissatisfaction. The anticipation of these strong emotions will dissuade individuals from not contributing, this may explain the emergence of the contribution

4 Discussion

In the model developed in the second section, players are expected to take into account the intentions of other players and react by maximizing their utilities by combining both gains and emotions. This integrated utility function of the relevant arguments, that takes into account their social or legal reasons and who actually refer to psychological considerations. In contrary to the conventional models, this consideration improves the explanatory power of the models with regard to experimental observations.

The experimental method allows precise and systematic measurement of real and controlled behaviors. This method makes it possible to isolate certain contexts, presented by theory as determinants of these behaviors, and to modify others in order to measure the effective impact. In particular, it makes it possible to exclude any possibility of contractual engagement, any reputation mechanism, and also any threat or promise.

Regarding the case studied in this article, the experiments has revealed several regularities. The first regularity is the observation that in the first period of the game, when individuals do not yet know the behavior of the other participants, on average, these individuals contribute about half of their endowment. The second pattern is the decline in average contribution levels between the first period and the last period of a repeated game (see, [16]). These observations highlight the concept of conditional cooperation and the importance of considering the heterogeneity of participant preferences to explain the tendency for contributions to decline over time.

Thus, according to these observations, consider that individuals contribute with half the dotation $c_k = 5$, then each player i has two choices: $c_i = 5$ or $c_i = 0$. In the first case its utility is $U = 10\lambda + 20\theta$.

If he is satisfied with his endowment and those of others, we will have $\lambda = \theta = 1$, and $U = 30$. On the other hand, if he chooses not to contribute $c_i = 0$, his utility is $U = 20\lambda + 15\theta + 15\omega$. By being not satisfied with his choice ($\lambda = 0$) and satisfied with the choices of others ($\theta = 1$) but anticipating a regret of maximum intensity $\omega = -1$, its utility will be $U = 0$. This means that the presence of the emotional feeling causes the utility of the non-contribution to be negligible, which leads the player to decide to contribute.

Not taking into account these maximum values of emotions will always give the chance to obtain $U(c_i = 5, c_k \neq 0) > U(c_i = 0, c_k \neq 0)$. This highlights the importance of conditional contribution, that is, the emotions that prevent the player from not contributing are present when the player anticipates that others will contribute, and they are weak otherwise. Thus, while respecting the principle of utility maximization, when the choices take into account the anticipated consequences of emotions, the outcome of the situation changes and consequently the equilibrium of the game changes.

Conditional cooperators are individuals who respect and enforce a social norm of cooperating when others cooperate. Therefore, the phenomenon of the decrease in cooperation over time stems from the conditionality of cooperation, which explains its emergence even in the absence of institutions (see, [17]). The importance of considering emotions in economic behavior simulating the triggering of a virtuous circle between sanction and cooperation (see, [16]).

Finally, because of the considerable interest of the treated subject, our research contributes to the debate on the financing of public goods and the question of the social preferences both on the theoretical and practical level.At the theoretical level, our research has highlighted the important role of emotions in the way of constructing utility functions in a particular case of public good by highlighting prosocial behavior. In addition, a better combination of earnings and emotions can describe and represent the material and emotional situation of a contributor. On the practical level, when agents are predisposed to contribute, public decision-makers on their part are invited to ensure the appropriate conditions in order to capitalize on this predisposition of citizens wishing to finance the public good. In addition, creating an environment that promotes trust and solidarity can be achieved through social recognition of the willingness to contribute by avoiding the stowaway paradox. It is about creating an economic and social environment likely to change cultures and mentalities for a new management of public goods which simulates social well-being.

5 Conclusion

Nowadays, the theory of social preferences is one of the theoretical innovations of economic analysis today. New models under this theory abandon the traditional image of the economic agent as a maximizer, calculator, who is preoccupied with his own material interest and selfish behavior. Theoretical models capturing social preferences are a very useful tool for economists who wish to rationalize empirical observations. They point to anomalies in the behavior of economic agents. From a standard economic analysis perspective, the economic agent is completely insensitive to emotions. In other words, the latter can in no way become involved in individual decisions. On the other hand, these new models abandon this image in favor of that of an actor subject to a wide range of emotions and feelings in his interactions with others.

The public good game illustrates the importance of embedding emotions in situations of social dilemma. It corresponds to situations in which each player makes his own choice of whether to contribute or not. Social dilemmas characterize in economics situations of social interaction where a conflict can arise between the interest of the individual and the social interest. These situations where the rationality of individuals prevents them from cooperating are traditionally illustrated by the prisoner's dilemma game.

Several facts are identified by experiential economics and which remain poorly explained by standard economic theory. Thus on the basis of these observations the authors have constructed models which account for these facts. Experiential research shows that individuals cooperate much more than predicted by classical economic theory based on individual selfishness.

Generally, research on social dilemmas encourages cooperation between researchers from different disciplines interested in understanding the sources of this evolution. Economists, managers, psychologists and sociologists can thus combine their efforts to understand the mechanisms behind the emergence of cooperation in social dilemmas (see, [16]).

References

1. Errays, N.A., Tourabi, A.: Le rôle du support du mari et de l'empathie dans la formation des intentions entrepreneuriales prosociales des femmes marocaines mariées. la revue gestion et organisation. **10**, 14–28 (2018)
2. Berg, J., Dickhaut, J., McCabe, K.: Trust, reciprocity, and history. Games Econ. Behav. **10**, 122–142 (1995)
3. Bolton, G., Ockenfels, A.: A theory of equity, reciprocity, and competition. Am. Econ. Rev. **90**, 166–193 (2000)
4. Charness, G., Rabin, M.: Understanding social preferences with simple tests. Q. J. Econ. **117**, 817–869 (2002)
5. Chaudhuri, A., Sopher, B., Strand, P.: Cooperation in social dilemmas, trust and reciprocity. J. Econ. Psychol. **23**, 231–249 (2002)
6. Dufwenberg, M., Kirchsteiger, G.: A theory of sequential reciprocity. Games Econ. Behav. **47**, 268–298 (2004)
7. Falk, A., Fischbacher, U.: A theory of reciprocity. Games Econ. Behav. **54**, 293–315 (2006)
8. Fehr, E., Gächter, S.: Cooperation and punishment in public goods experiments. Am. Econ. Rev. **66**(2), 980–994 (2000)
9. Fehr, E., Schmidt, K.: A theory of fairness, competition, and cooperation. Q. J. Econ. **114**, 817–868 (1999)
10. Fischbacher, U., Gächter, S., Fehr, E.: Are people conditionally cooperative? Evidence from a public goods experiment. Econ. Lett. **71**, 397–404 (2001)
11. Fischbacher, U., Gächter, S.: Social preferences, beliefs, and the dynamics of freeriding in public good experiments. Am. Econ. Rev. **100**(1), 541–556 (2010)
12. Isaac, M., Walker, J.: communication and free-riding behavior: the voluntary contributions mechanism. Econ. Inq. **26**(4), 585–608 (1988)
13. Isaac, M., Walker, J., Williams, A.: Group size and voluntary provision of public goods: experimental evidence utilizing large groups. J. Public Econ. **54**, 1–36 (1994)
14. Jourdheuil, R., Petit, P.: Émotions morales et comportement prosocial: Une revue de la littérature. Revue d'Économie Politique **4**(125), 499–525 (2015)
15. Ledyard J.: Public goods: a survey of experimental research. In: Kagel, J., Roth, A. (eds.) Handbook of Experimental Economics, pp. 111–194. Princeton, Princeton university Press (1995)
16. Villeval, M.C.: Quand le marché ne suffit plus: biens publics et coopèration conditionnelle. Idées économiques et sociales 2010/3 N 161, 6–14 (2010). https://www.cairn.info/revue-idees-economiques-et-sociales-2010-3-page-6.htm [Consulté le 28 octobre 2020]
17. Villeval, M.C.: Contribution aux biens publics et préférences sociales - Apports récents de l'économie comportementale. Revue économique, Vol. 3 N 161, pp. 6–14 (2012). https://www.cairn.info/revue-economique-2012-3-page-389.htm [Consulté le 26 octobre 2020]
18. Rabin, M.: Incorporating fairness into game theory and economics. Am. Econ. Rev. **80**(5), 1281–1302 (1993)

Fundamental Systems of Units of Some Imaginary Multiquadratic Fields of Degree 16

Abdelmalek Azizi, Mohamed Mahmoud Chems-Eddin, and Abdelkader Zekhnini

Abstract Let $q_1 \equiv q_2 \equiv 3 \pmod 8$ be two different prime integers, d a positive odd square-free integer relatively prime to q_1 and q_2. The main aim of this paper is to investigate the unit groups of some number fields of the form $\mathbb{L} = \mathbb{Q}(\sqrt{2}, \sqrt{q_1}, \sqrt{q_2}, \sqrt{-d})$.

Keywords Multiquadratic number fields · Unit group · Unit index · Quadratic fields

1 Introduction

Let k be a number field of degree n, (i.e., $[k : \mathbb{Q}] = n$). Denote by E_k the unit group of k that is the group of the invertible elements of the ring $\mathcal{O}_k$ of algebraic integers of the number field k. By the well known Dirichlet's unit theorem, if $n = r_1 + 2r_2$, where r_1 is the number of real embeddings and r_2 the number of conjugate pairs of complex embeddings of k, then there exist $r = r_1 + r_2 - 1$ units of $\mathcal{O}_k$ that generate E_k (modulo the roots of unity), and these r units are called the *fundamental system of units of k*. Therefore

$$E_k \simeq \mu(k) \times \mathbb{Z}^{r_1+r_2-1},$$

where $\mu(k)$ is the group of roots of unity contained in k.

One major problem in algebraic number theory (and thus in theory of units of number fields which is related to all areas of algebraic number theory) is the computation of the fundamental system of units. For quadratic fields, the problem is easily solved. For quartic bicyclic fields, Kubota [6] gave a method for finding a fundamen-

A. Azizi · A. Zekhnini
Mathematics Department, Sciences Faculty, Mohammed Premier University, Oujda, Morocco

M. M. Chems-Eddin (✉)
Department of Mathematics, Faculty of Sciences Dhar El Mahraz, University Sidi Mohamed Ben Abdellah, Fez, Morocco
e-mail: 2m.chemseddin@gmail.com

S. Melliani et al. (eds.), *Applied Mathematics and Modelling in Finance, Marketing and Economics*, Studies in Computational Intelligence 1114,
https://doi.org/10.1007/978-3-031-42847-0_11

tal system of units. Wada [7] generalized Kubota's method, creating an algorithm for computing fundamental units in any given multiquadratic field. However, in general, it is not easy to compute the unit group of a number field especially for number fields of degree more than 4. Actually, in literature there are only few examples of computation of unit group of a given number field k of degree 16 (see our recent works [2–4]).

In this paper, the main goal is to determine the $r = r_1 + r_2 - 1$ generators of the torsion-free subgroup of E_k for an infinite family of number fields k of degree 16 of the form $\mathbb{Q}(\sqrt{2}, \sqrt{q_1}, \sqrt{q_2}, \sqrt{-d})$, where $q_1 \equiv q_2 \equiv 3 \pmod 8$ are two different prime integers and d a positive odd square-free integer. Let ε_ℓ denote the fundamental unit of the quadratic field $\mathbb{Q}(\sqrt{\ell})$ and $\left(\frac{\cdot}{\cdot}\right)$ the Legendre Symbol. Then the main theorem of this paper is the following.

Theorem 1 *Let $q_1 \equiv q_2 \equiv 3 \pmod 8$ be two different prime integers, d a positive odd square-free integer relatively prime to q_1 and q_2, and $\mathbb{L} = \mathbb{Q}(\sqrt{2}, \sqrt{q_1}, \sqrt{q_2}, \sqrt{-d})$. Without loss of generality we may assume that $\left(\frac{q_1}{q_2}\right) = 1$. So we have:*

1. *If $d = 1$, then a fundamental system of units of $\mathbb{L}$ is given by*

$$\left\{\varepsilon_2, \sqrt{\varepsilon_{q_1}}, \sqrt{\varepsilon_{q_2}}, \sqrt{\varepsilon_{q_1q_2}}, \sqrt{\sqrt{\varepsilon_{q_1}}\sqrt{\varepsilon_{q_2}}\sqrt{\varepsilon_{2q_1q_2}}}, \sqrt{\sqrt{\varepsilon_{2q_1}}\sqrt{\varepsilon_{2q_2}}\sqrt{\varepsilon_{2q_1q_2}}}, \sqrt{\zeta_8\varepsilon_2\sqrt{\varepsilon_{q_1}}\sqrt{\varepsilon_{2q_1}}}\right\},$$

   where ζ_8 is the primitive 8-th root of unity.
2. *If $d \neq 1$, then a fundamental system of units of $\mathbb{L}$ is given by*

$$\left\{\varepsilon_2, \sqrt{\varepsilon_{q_1}}, \sqrt{\varepsilon_{2q_1}}, \sqrt{\varepsilon_{q_2}}, \sqrt{\varepsilon_{q_1q_2}}, \sqrt{\sqrt{\varepsilon_{q_1}}\sqrt{\varepsilon_{q_2}}\sqrt{\varepsilon_{2q_1q_2}}}, \sqrt{\sqrt{\varepsilon_{2q_1}}\sqrt{\varepsilon_{2q_2}}\sqrt{\varepsilon_{2q_1q_2}}}\right\}.$$

The proof this Theorem is long and very technical. Therefore, in the third section, we prove with details the first item of the main Theorem and we shall let details of the proof of the second item to the reader.

2 Preliminary Results

In this section we recall some results that will be useful in what follows.

Lemma 1 *Let K_0 be a real number field, $K = K_0(i)$ a quadratic extension of K_0, $n \geq 2$ an integer and ξ_n a 2^n-th primitive root of unity, then $\xi_n = \frac{1}{2}(\mu_n + \lambda_n i)$, where $\mu_n = \sqrt{2 + \mu_n}$, $\lambda_n = \sqrt{2 - \mu_{n-1}}$, $\mu_2 = 0$, $\lambda_2 = 2$ and $\mu_3 = \lambda_3 = \sqrt{2}$. Let n_0 be the greatest integer such that ξ_{n_0} is contained in K, $\{\epsilon_1, \ldots, \epsilon_r\}$ a fundamental system of units of K_0 and ϵ a unit of K_0 such that $(2 + \mu_{n_0})\epsilon$ is a square in K_0 (if it exists). Then a fundamental system of units of K is one of the following systems:*

1. *$\{\epsilon_1, \ldots, \epsilon_{r-1}, \sqrt{\xi_{n_0}\epsilon}\}$ if ϵ exists, in this case $\epsilon = \epsilon_1^{j_1} \cdots \epsilon_{r-1}^{j_1}\epsilon_r$, where $j_i \in \{0, 1\}$.*
2. *$\{\epsilon_1, \ldots, \epsilon_r\}$ elsewhere.*

Proof See [1, Proposition 2].

Lemma 2 *Let $K_0/\mathbb{Q}$ be an abelian extension such that K_0 is real and β a positive square-free algebraic integer of K_0. Assume that $K = K_0(\sqrt{-\beta})$ is a quadratic extension of K_0, which is abelian over $\mathbb{Q}$. Assume furthermore that $i = \sqrt{-1} \notin K$. Let $\{\epsilon_1, \ldots, \epsilon_r\}$ be a fundamental system of unit of K_0. Without loss of generality we may suppose that the units ϵ_i are positives. Let ϵ be a unit of K_0 such that $\beta\epsilon$ is a square in K_0 (if it exists). Then a fundamental system of units of K is one of the following systems:*

1. $\{\epsilon_1, \ldots, \epsilon_{r-1}, \sqrt{-\epsilon}\}$ *if ϵ exists, in this case $\epsilon = \epsilon_1^{j_1} \cdots \epsilon_{r-1}^{j_1}\epsilon_r$, where $j_i \in \{0, 1\}$.*
2. $\{\epsilon_1, \ldots, \epsilon_r\}$ *else.*

Proof See [1, Proposition 3].

Lemma 3 *Let $q_1 \equiv q_2 \equiv 3 \pmod 8$ be two primes such that $\left(\frac{q_1}{q_2}\right) = 1$.*

1. *Let x and y be two integers such that $\varepsilon_{2q_1q_2} = x + y\sqrt{2q_1q_2}$. Then*
   a. *$x - 1$ is a square in $\mathbb{N}$,*
   b. *$\sqrt{2\varepsilon_{2q_1q_2}} = y_1 + y_2\sqrt{2q_1q_2}$ and $2 = -y_1^2 + 2q_1q_2y_2^2$, for some integers y_1 and y_2 satisfying $y = y_1y_2$.*
2. *There are two integers a and b such that $\varepsilon_{q_1q_2} = a + b\sqrt{q_1q_2}$ and we have*
   a. *$2q_1(a + 1)$ is a square in $\mathbb{N}$,*
   b. *b is even, $\sqrt{\varepsilon_{q_1q_2}} = b_1\sqrt{q_1} + b_2\sqrt{q_2}$ and $1 = q_1b_1^2 - q_2b_2^2$ for some integers b_1 and b_2 such that $b = 2b_1b_2$.*
3. *Let c and d be two integers such that $\varepsilon_{2q_i} = c + d\sqrt{2q_i}$. Then we have*
   a. *$c - 1$ is a square in $\mathbb{N}$,*
   b. *$\sqrt{2\varepsilon_{2q_i}} = d_1 + d_2\sqrt{2q_i}$ and $2 = -d_1^2 + 2q_id_2^2$, for some integers d_1 and d_2 such that $d = d_1d_2$.*
4. *Let α and β be two integers such that $\varepsilon_{q_i} = \alpha + \beta\sqrt{q_i}$. Then we have*
   a. *$\alpha - 1$ is a square in $\mathbb{N}$,*
   b. *$\sqrt{2\varepsilon_{q_i}} = \beta_1 + \beta_2\sqrt{q_i}$ and $2 = -\beta_1^2 + q_i\beta_2^2$, for some integers β_1 and β_2 such that $\beta = \beta_1\beta_2$.*

Proof See [5, Lemma 2.4].

3 Proof of the Main Theorem

Now we can prove our main theorem. Let us start by the first item:

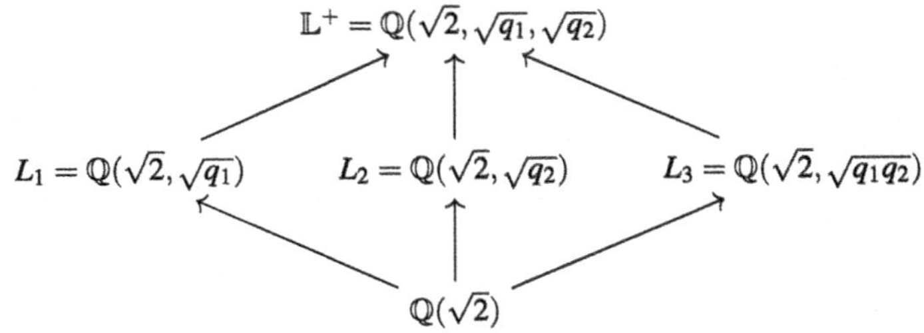

Fig. 1 Subfields of $\mathbb{L}^+/\mathbb{Q}(\sqrt{2})$

1. Without loss of generality we may suppose that $\left(\frac{q_1}{q_2}\right) = 1$. First we shell need the fundamental system of units of $\mathbb{L}^+ = \mathbb{Q}(\sqrt{2}, \sqrt{q_1}, \sqrt{q_2})$, and then using Lemma 1 we deduce the fundamental system of units of $\mathbb{L}$. Consider the following diagram (Fig. 1 below).
Put $\mathrm{Gal}(\mathbb{L}^+/\mathbb{Q}) = \langle \sigma_1, \sigma_2, \sigma_3 \rangle$, where

$$\begin{array}{lll} \sigma_1(\sqrt{2}) = -\sqrt{2}, & \sigma_1(\sqrt{q_1}) = \sqrt{q_1}, & \sigma_1(\sqrt{q_2}) = \sqrt{q_2} \\ \sigma_2(\sqrt{2}) = \sqrt{2}, & \sigma_2(\sqrt{q_1}) = -\sqrt{q_1}, & \sigma_2(\sqrt{q_2}) = \sqrt{q_2} \\ \sigma_3(\sqrt{2}) = \sqrt{2}, & \sigma_3(\sqrt{q_1}) = \sqrt{q_1}, & \sigma_3(\sqrt{q_2}) = -\sqrt{q_2}. \end{array}$$

By [5, Proposition 2.7], we have

$$E_{\mathbb{L}^+} = \left\langle -1, \varepsilon_2, \sqrt{\varepsilon_{q_1}}, \sqrt{\varepsilon_{2q_1}}, \sqrt{\varepsilon_{q_2}}, \sqrt{\varepsilon_{q_1q_2}}, \sqrt{\sqrt{\varepsilon_{q_1}}\sqrt{\varepsilon_{q_2}}\sqrt{\varepsilon_{2q_1q_2}}}, \sqrt{\sqrt{\varepsilon_{2q_1}}\sqrt{\varepsilon_{2q_2}}\sqrt{\varepsilon_{2q_1q_2}}} \right\rangle.$$

Put

$$\xi^2 = (2+\sqrt{2}) \cdot \varepsilon_2^a \cdot \sqrt{\varepsilon_{q_1}}^b \cdot \sqrt{\varepsilon_{2q_1}}^c \cdot \sqrt{\varepsilon_{q_2}}^d \cdot \sqrt{\varepsilon_{q_1q_2}}^e \cdot \sqrt[4]{\varepsilon_{q_1}\varepsilon_{q_2}\varepsilon_{2q_1q_2}}^f \cdot \sqrt[4]{\varepsilon_{2q_1}\varepsilon_{2q_2}\varepsilon_{2q_1q_2}}^g,$$

with $a, b, c, d, e, f, g \in \{0, 1\}$. We will use norm maps from $\mathbb{L}^+$ to its biquadratic subextensions. The computations of these norms are summarized in the following table (see Table 1). Note that the third line of Table 1, is constructed as follows (we similarly construct the rest of the table):
By Lemma 3, we have $\sqrt{\varepsilon_{q_1}} = \frac{1}{\sqrt{2}}(\beta_1 + \beta_2\sqrt{q_1})$ and $2 = -\beta_1^2 + q_1\beta_2^2$. Thus:

$$\begin{aligned} \sqrt{\varepsilon_{q_1}}^{\sigma_1} &= \frac{1}{-\sqrt{2}}(\beta_1 + \beta_2\sqrt{q_1}) \\ &= -\sqrt{\varepsilon_{q_1}}. \end{aligned}$$

$$\begin{aligned}\sqrt{\varepsilon_{q_1}}^{\sigma_2} &= \tfrac{1}{\sqrt{2}}(\beta_1 - \beta_2\sqrt{q_1})\\ &= \tfrac{1}{\sqrt{2}} \cdot \tfrac{(\beta_1-\beta_2\sqrt{q_1})(\beta_1+\beta_2\sqrt{q_1})}{\beta_1+\beta_2\sqrt{q_1}}\\ &= \tfrac{1}{\sqrt{2}} \cdot \tfrac{(\beta_1^2-\beta_2^2 q_1)}{\sqrt{2}\sqrt{\varepsilon_{q_1}}}\\ &= \tfrac{1}{2} \cdot \tfrac{-2}{\sqrt{\varepsilon_{q_1}}} = \tfrac{-1}{\sqrt{\varepsilon_{q_1}}}.\end{aligned}$$

$$\begin{aligned}\sqrt{\varepsilon_{q_1}}^{\sigma_3} &= \tfrac{1}{\sqrt{2}}(\beta_1 + \beta_2\sqrt{q_1})\\ &= \sqrt{\varepsilon_{q_1}}.\end{aligned}$$

$$\begin{aligned}\sqrt{\varepsilon_{q_1}}^{1+\sigma_1} &= \sqrt{\varepsilon_{q_1}} \cdot \sigma_1(\sqrt{\varepsilon_{q_1}})\\ &= \sqrt{\varepsilon_{q_1}} \cdot (-\sqrt{\varepsilon_{q_1}})\\ &= -\varepsilon_{q_1}.\end{aligned}$$

$$\begin{aligned}\sqrt{\varepsilon_{q_1}}^{1+\sigma_2} &= \sqrt{\varepsilon_{q_1}} \cdot \sigma_2(\sqrt{\varepsilon_{q_1}})\\ &= \sqrt{\varepsilon_{q_1}} \cdot (\tfrac{-1}{\sqrt{\varepsilon_{q_1}}})\\ &= -1.\end{aligned}$$

$$\begin{aligned}\sqrt{\varepsilon_{q_1}}^{1+\sigma_1\sigma_3} &= \sqrt{\varepsilon_{q_1}} \cdot \sigma_1(\sigma_3(\sqrt{\varepsilon_{q_1}}))\\ &= \sqrt{\varepsilon_{q_1}} \cdot \sigma_1(\sqrt{\varepsilon_{q_1}})\\ &= \sqrt{\varepsilon_{q_1}} \cdot (-\sqrt{\varepsilon_{q_1}})\\ &= -\varepsilon_{q_1}.\end{aligned}$$

$$\begin{aligned}\sqrt{\varepsilon_{q_1}}^{1+\sigma_2\sigma_3} &= \sqrt{\varepsilon_{q_1}} \cdot \sigma_2(\sigma_3(\sqrt{\varepsilon_{q_1}}))\\ &= \sqrt{\varepsilon_{q_1}} \cdot \sigma_2(\sqrt{\varepsilon_{q_1}})\\ &= \sqrt{\varepsilon_{q_1}} \cdot \tfrac{-1}{\sqrt{\varepsilon_{q_1}}}\\ &= -1.\end{aligned}$$

• Let us eliminate some forms of ξ^2 such that ξ can not be in $\mathbb{L}$. Consider $L_4 = \mathbb{Q}(\sqrt{q_1}, \sqrt{q_2})$, we will apply the norm $N_{\mathbb{L}/L_4} = 1 + \sigma_1$.

$$\begin{aligned}N_{\mathbb{L}/L_4}(\xi^2) &= 2 \cdot (-1)^a \cdot (-1)^b \cdot \varepsilon_{q_1}^b \cdot 1 \cdot (-1)^d \cdot (\varepsilon_{q_2})^d \cdot \varepsilon_{q_1q_2}^e \cdot (-1)^{uf} \sqrt{\varepsilon_{q_1}}^f \sqrt{\varepsilon_{q_2}}^f \cdot (-1)^{gv},\\ &= (-1)^{a+b+d+uf+gv} \cdot 2 \cdot \varepsilon_{q_1}^b \cdot \varepsilon_{q_2}^d \cdot \varepsilon_{q_1q_2}^e \cdot \sqrt{\varepsilon_{q_1}}^f \cdot \sqrt{\varepsilon_{q_2}}^f.\end{aligned}$$

Therefore, $a + b + d + uf + gv \equiv 0 \pmod 2$. One can easily deduce that $f = 0$. Thus $a + b + d + gv \equiv 0 \pmod 2$ and

$$\xi^2 = (2 + \sqrt{2}) \cdot \varepsilon_2^a \cdot \sqrt{\varepsilon_{q_1}}^b \cdot \sqrt{\varepsilon_{2q_1}}^c \cdot \sqrt{\varepsilon_{q_2}}^d \cdot \sqrt{\varepsilon_{q_1q_2}}^e \cdot \sqrt[4]{\varepsilon_{2q_1}\varepsilon_{2q_2}\varepsilon_{2q_1q_2}}^g.$$

Table 1 Norms in $\mathbb{L}^+/\mathbb{Q}(\sqrt{2})$

ε	ε^{σ_1}	ε^{σ_2}	ε^{σ_3}	$\varepsilon^{1+\sigma_1}$	$\varepsilon^{1+\sigma_2}$	$\varepsilon^{1+\sigma_1\sigma_3}$	$\varepsilon^{1+\sigma_2\sigma_3}$
ε_2	$\frac{-1}{\sqrt{\varepsilon_2}}$	ε_2	ε_2	-1	ε_2^2	-1	ε_2^2
$\sqrt{\varepsilon_{q_1}}$	$-\sqrt{\varepsilon_{q_1}}$	$\frac{-1}{\sqrt{\varepsilon_{q_1}}}$	$\sqrt{\varepsilon_{q_1}}$	$-\varepsilon_{q_1}$	-1	$-\varepsilon_{q_1}$	-1
$\sqrt{\varepsilon_{2q_1}}$	$\frac{1}{\sqrt{\varepsilon_{2q_1}}}$	$\frac{-1}{\sqrt{\varepsilon_{2q_1}}}$	$\sqrt{\varepsilon_{2q_1}}$	1	-1	1	-1
$\sqrt{\varepsilon_{q_2}}$	$-\sqrt{\varepsilon_{q_2}}$	$\sqrt{\varepsilon_{q_2}}$	$\frac{-1}{\sqrt{\varepsilon_{q_2}}}$	$-\varepsilon_{q_2}$	ε_{q_2}	1	-1
$\sqrt{\varepsilon_{2q_2}}$	$\frac{1}{\sqrt{\varepsilon_{2q_2}}}$	$\sqrt{\varepsilon_{2q_2}}$	$\frac{-1}{\sqrt{\varepsilon_{2q_2}}}$	1	ε_{2q_2}	$-\varepsilon_{2q_2}$	-1
$\sqrt{\varepsilon_{q_1q_2}}$	$\sqrt{\varepsilon_{q_1q_2}}$	$\frac{-1}{\sqrt{\varepsilon_{q_1q_2}}}$	$\frac{1}{\sqrt{\varepsilon_{q_1q_2}}}$	$\varepsilon_{q_1q_2}$	-1	1	$-\varepsilon_{q_1q_2}$
$\sqrt{\varepsilon_{2q_1q_2}}$	$\frac{1}{\sqrt{\varepsilon_{2q_1q_2}}}$	$\frac{-1}{\sqrt{\varepsilon_{2q_1q_2}}}$	$\frac{-1}{\sqrt{\varepsilon_{2q_1q_2}}}$	1	-1	$-\varepsilon_{2q_1q_2}$	$\varepsilon_{2q_1q_2}$

• Now we will apply the norm $N_{\mathbb{L}/L_3} = 1 + \sigma_2\sigma_3$, where $L_3 = \mathbb{Q}(\sqrt{2}, \sqrt{q_1q_2})$. We have

$$N_{\mathbb{L}/L_3}(\xi^2) = (2+\sqrt{2})^2 \cdot \varepsilon_2^{2a} \cdot (-1)^b \cdot (-1)^c \cdot (-1)^d \cdot (-1)^e \cdot \varepsilon_{q_1q_2}^e \cdot (-1)^{tg} \cdot \sqrt{\varepsilon_{2q_1q_2}}^g,$$
$$= (2+\sqrt{2})^2 \cdot \varepsilon_2^{2a} \cdot (-1)^{b+c+d+e+tg} \cdot \varepsilon_{q_1q_2}^e \cdot \sqrt{\varepsilon_{2q_1q_2}}^g.$$

Using Lemma 3, it is easy to deduce that $e = g = 0$. Thus $b + c + d \equiv 0 \pmod 2$ and $a + b + d \equiv 0 \pmod 2$. It follows that $a = c$ and

$$\xi^2 = (2+\sqrt{2}) \cdot \varepsilon_2^a \cdot \sqrt{\varepsilon_{q_1}}^b \cdot \sqrt{\varepsilon_{2q_1}}^a \cdot \sqrt{\varepsilon_{q_2}}^d.$$

• Let us apply $N_{\mathbb{L}/L_5} = 1 + \sigma_1\sigma_3$, with $L_5 = \mathbb{Q}(\sqrt{q_1}, \sqrt{2q_2})$. We have

$$N_{\mathbb{L}/L_3}(\xi^2) = 2 \cdot (-1)^a \cdot (-1)^b \cdot \varepsilon_{q_1}^b \cdot 1 \cdot 1$$
$$= (-1)^{a+b} \cdot 2 \cdot \varepsilon_{q_1}^b.$$

So $a + b \equiv 0 \pmod 2$. Since 2 is not a square in L_5, then using Lemma 3, one easily deduces that $b = 1$ and so $a = 1$. Since $a + b + d \equiv 0 \pmod 2$, then $d = 0$. Therefore,

$$\xi^2 = (2+\sqrt{2}) \cdot \varepsilon_2 \cdot \sqrt{\varepsilon_{q_1}} \cdot \sqrt{\varepsilon_{2q_1}}$$

Since the Hasse's unit index $Q_{\mathbb{L}}$ equals 2 (cf. the proof of the main theorem of [5]), then by Lemma 1 $(2+\sqrt{2}) \cdot \varepsilon_2 \cdot \sqrt{\varepsilon_{q_1}} \cdot \sqrt{\varepsilon_{2q_1}}$ is a square and therefore we have the first item.

2. For the proof of the second item we similarly put

$$\xi^2 = d \cdot \varepsilon_2^a \cdot \sqrt{\varepsilon_{q_1}}^b \cdot \sqrt{\varepsilon_{2q_1}}^c \cdot \sqrt{\varepsilon_{q_2}}^d \cdot \sqrt{\varepsilon_{q_1q_2}}^e \cdot \sqrt[4]{\varepsilon_{q_1}\varepsilon_{q_2}\varepsilon_{2q_1q_2}}^f \cdot \sqrt[4]{\varepsilon_{2q_1}\varepsilon_{2q_2}\varepsilon_{2q_1q_2}}^g,$$

with $a, b, c, d, e, f \in \{0, 1\}$. We proceed as above to eliminate all forms of ξ^2, and we deduce the result by using Lemma 2.

References

1. Azizi, A.: Unités de certains corps de nombres imaginaires et abéliens sur $\mathbb{Q}$. Ann. Sci. Math. Québec. **23**, 15–21 (1999)
2. Chems-Eddin, M.M.: Unit groups of some multiquadratic number fields and 2-class groups. Period. Math. Hung. **84**, 235–249 (2022)
3. Chems-Eddin, M.M., Azizi, A., Zekhnini, A.: Unit groups and Iwasawa lambda invariants of some multiquadratic number fields. Bol. Soc. Mat. Mex. III. Ser. **27** (2021), Article ID 24, 16 pages
4. Chems-Eddin, M.M., Zekhnini, A., Azizi, A.: Units and 2-class field towers of some multiquadratic number fields. Turk. J. Math. **44**, 1466–1483 (2020)
5. Chems-Eddin, M.M., Zekhnini, A., Azizi, A.: On the Hilbert 2-class field towers of some cyclotomic $\mathbb{Z}_2$-extensions. arXiv:2005.06646
6. Kubota, T.: Über den bizyklischen biquadratischen Zahlkörper. Nagoya Math. J. **10**, 65–85 (1956) (in German)
7. Wada, H.: On the class number and the unit group of certain algebraic number fields. J. Fac. Univ. Tokyo. **13**, 201–209 (1966)

One-Dimensional Inverse Stefan Problem Numerical Approximation Utilizing a Meshless Method

Mohammed Baati and Mohamed Louzar

Abstract We extend a meshless method of fundamental solutions to the one-dimensional inverse Stefan problem for the heat equation, where the boundary data is to be reconstructed on the fixed boundary. The inverse problem is ill-posed for small errors in the input measured data can cause high deviations in solution. Therefore, we incorporate Tikhonov regularization to obtain a stable solution. Numerical results are presented.

1 Introduction

Stefan problems model is a specific type of free boundary problems in partial differential equations related to heat diffusion [1, 2]. It aims to describe the temperature distribution in a homogeneous medium, for examples solidification of metals, freezing of water and food, melting, ablation, etc.

The direct Stefan problem consists finding the temperature and the moving boundary interface when the boundary conditions and the initials, and the thermal properties of the heat-conducting body are known [3]. The inverse problem consists a calculation of temperature distribution, as good as the reconstruction of the function which describes the temperature distribution on the boundary and/or the initial conditions, and/or thermal properties from additional information, which may involve the partial knowledge or measurement of the moving boundary interface position, its velocity in a normal direction, or the temperature at selected interior points of the domain, examples of inverse solidification problems can be found in [1, 4].

Recently, the Fundamental Solutions Method (MFS) has been successfully applied to different types of hyperbolic and parabolic problems, including direct and inverse Stefan problems. The simplicity of the method and the ease with which it can be implemented have made it very well known.

M. Baati (✉) · M. Louzar
Laboratory MISI, Faculty of Science and Technology of Settat, University Hassan 1 er, Settat, Morocco
e-mail: mohammedbaati11@gmail.com

S. Melliani et al. (eds.), *Applied Mathematics and Modelling in Finance, Marketing and Economics*, Studies in Computational Intelligence 1114,
https://doi.org/10.1007/978-3-031-42847-0_12

For the advantages of MFS in comparison to other more classical methods of domain, we can refer to [5]. For the outline of this paper in Sect. 2 we formulate the inverse Stefan problem and discuss some variants of it. In Sect. 3 we give an MFS for numerically solving the inverse Stefan problem and recall some theoretical properties about the denseness of this MFS approximation. In Sect. 4 the numerical results presented show that an accurate approximation to the inverse Stefan problem can be obtained using the MFS with short computational cost.

2 Mathematical Formulation

Consider the direct one-dimensional Stefan problem, we like to determine the free boundary (sufficiently smooth) which we denote by Rs and is given by x = s(t) for t ∈ (0, T] and the temperature solution u(x, t). Let D = (0, $s(t)$)× (0, T] denote the heat conduction domain with $\bar{D}$ = [0, s(t)] × [0, T] and For any positive function s ∈ C[0, T] satisfying s(0) = b, where T > 0 and b ≥ 0 be two fixed constants.

We have a fixed boundary at x = 0, which we denote by R_u. We denote the union of the boundaries by R = $R_S \cup R_u$.

In the direct one-phase Stefan problem we wish to determine the solution u(x, t) as well as the moving boundary given by x = s(t), satisfying the heat equation:

$$\frac{\partial u}{\partial t}(x,t) - \frac{\partial^2 u}{\partial x^2}(x,t) = 0, (x,t) \in D \tag{1}$$

Subject to the initial condition:

$$u(x,0) = u_0(x), x \in (0,b], s(0) = b \tag{2}$$

The Dirichlet and Neumann boundary conditions on the moving boundary x = s(t)

$$u(s(t),t) = 0, t \in (0,T] \tag{3}$$

$$\frac{\partial u}{\partial x}(s(t),t) = -s', t \in (0,T] \tag{4}$$

At the fixed boundary x = 0, we have the following Dirichlet and Neumann data:

$$u(0,t) = h(t), t \in (0,T] \tag{5}$$

$$-\frac{\partial u}{\partial x}(s(t),t) = g(t), t \in (0,T] \tag{6}$$

Existence, uniqueness of a solution and continuous dependence on the data, i.e. wellposedness for u(x, t) and s(t), hold for direct one-phase Stefan problem (1)–(5), (1)–(4) and (6), see [6–8].

The boundary conditions given by Eqs. (3) and (4) can be changed by the general boundary conditions

$$u(s(t), t) = h_1(t), t \in (0, T] \tag{7}$$

$$\frac{\partial u}{\partial x}(s(t), t) = h_2(t), t \in (0, T] \tag{8}$$

where $h_1, h_2 \in C_1([0, T])$ satisfy the compatibility conditions $h_2(0) = u_0(s(0))$ and $h_1(0) = u_0'(s(0))$.

The inverse Stefan problem that will be investigated in this paper needs to finding the temperature u(x, t) satisfying Eqs. (1)–(4) with s(t) prescribed and reconstructing the Dirichlet and Neumann data at the fixed boundary at x = 0. This inverse Stefan problem is still an ill-posed problem with respect to short perturbations of the input data. Thus, regularization techniques are needed to restore stability.

3 The Method of Fundamental Solution (MFS)

We recall the fundamental solution of the one-dimensional heat Eq. (1):

$$F(x, t, y, \tau) = \frac{H(t-\tau)}{(4\pi(t-\tau))^{\frac{1}{2}}} e^{\frac{-(x-y)^2}{4(t-\tau)}} \tag{9}$$

where H is the Heaviside function.

We make a set of source points placed external to the domain $\bar{D}$. We indicate the domain by D_E, which contains the domain $\bar{D}$, with bounding surface R_E, and consider the time interval (0, T) extended (−T, T), see Fig. 1 for a representation of the domain, boundary and placement of source points. The boundary R_E is split into two boundaries R_E^1 (for x < 0 with points or this boundary denoted by y1(t)) and R_E^2) (for x > s(t) with points on this boundary indicated by y2(t)).

We make a denumerable, wherever dense set of source points on the external boundary R_E, and denote this set by $(y_j(\tau_m), \tau_m)_{1,2\dots}$ and $j = 1, 2$ be a denumerable, wherever dense set of source points equally distributed on the external boundary R_E. The fundamental solution verify the heat equation:

$$\frac{\partial F}{\partial t} - \frac{\partial^2 F}{\partial x^2} = 0 \tag{10}$$

We make an method of fundamental solution approximation to (1)–(4) as a linear combination of these fundamental solution given by

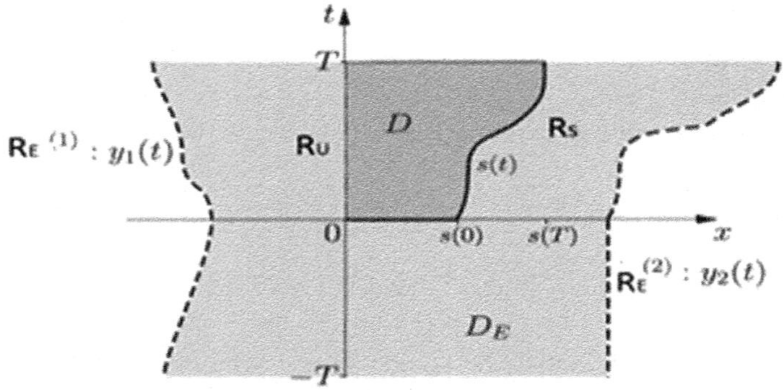

Fig. 1 General representation of the domain D and boundary $R = R_u \cup R_s$, with unspecified boundary condition (...) R_u and source points (—) placed on $R_E = R_E^1 \cup R_E^2$ external to the domain D

$$u_\infty(x,t) = \sum_{j=1}^{2}\sum_{m=1}^{\infty} c_m^j F(x,t; y_j(\tau_m), \tau_m), (x,t) \in \bar{D} \tag{11}$$

Theorem 1 *Elements of the form (11) constitute a linear independent and dense set in the L^2-sense on the initial base boundary as well as on the boundaries R_u and R_E.*

To implement the MFS for the inverse Stefan problem we make truncate (11) by taking a finite number of terms, namely

$$u_M(x,t) = \sum_{j=1}^{2}\sum_{m=1}^{2M} c_m^j F(x,t; y_j(\tau_m), \tau_m), (x,t) \in \bar{D} \tag{12}$$

We construct a one-dimensional domain, with a fixed boundary at x = 0 and a moving boundary $x = s(t)$, and source points placed at the coordinates following:

$$(-h, \tau) \ for \ \tau \in (-T, T) \tag{13}$$

$$(h + s(\tau), \tau) \ for \ \tau \in (0, T), and \ (h + s(-\tau), \tau) \ for \ \tau \in (-T, 0) \tag{14}$$

Source points have been settled symmetrically with respect to τ via a reflection through t = 0, we will test other source point positions to see if a different placement produces better results.

The source points will be settled at time points $(\tau_m)_{m=1,\dots,2M} \in (-T, T)$ given by

$$\tau_m = \frac{2(m-M)-1}{2M} T, m = 1, \dots, 2M \tag{15}$$

and on R_E set

$$y_1(\tau_m) = -h \; and \; y_2(\tau_m) = h + s(|\tau_m|), m = 1, \ldots, 2M \tag{16}$$

we have in total 4M source points on the external boundary R_E, and we make the same number of collocation points on the lateral and base surfaces $S_0 \cup R_s$. We indicate that the position of the collocation points may be arbitrary, but in what follows we place them for ease of implementation. Let

$$t_i = \frac{i}{M}T, x_1^i = s(t_i), i = 0, \ldots, M, \tag{17}$$

$$x_0^k = \frac{ks(0)}{K+1}, k = 1, \ldots, K \tag{18}$$

we construct the following system of equations

$$u_M(x_0^k, 0) = u_0(x_0^k), k = 1, \ldots, K \tag{19}$$

$$u_M(x_1^i, t_i) = 0, \tag{20}$$

$$\frac{\partial u_M}{\partial x}(x_1^i, t_i) = -s'(t_i), i = 0, \ldots, M \tag{21}$$

The system of Eqs. (19)–(21) contains $k + 2(M + 1)$ equations and $4M$ unknowns. Therefore, a necessary condition for a unique solution is $K \geq 2M - 2$. This system can be represented by

$$Ac = g \tag{22}$$

where c indicates the vector of unknown constants c_m^j, g represent the initial and boundary values at the collocation points and A is the matrix which indicate the value of fundamental solution at the corresponding collocation and source points. The matrix A will have a high condition number, so regularization is usually required. We make the Tikhonov regularization method, which solves the modified system of equations following:

$$(A^{tr}A + \lambda I)c = A^{tr}g \tag{23}$$

instead of the system of Eq. (22), where the superscript ^{tr} indicates the transpose of a matrix and I is the identity matrix. The now well-conditioned linear system of Eq. (23) can be solved using any classical method such as the Gaussian elimination method for example. The regularization parameter $\lambda \geq 0$ is taken according to the L-curve criterion [7]. We indicate that solving the system of Eq. (23) gives an approximation to the vector of coefficients c.

4 Numerical Results

We have presented an example of a test in this section such that s(t) is a linear function [8].

$$s(t) = t + b, t \in [0, T = 1] \tag{24}$$

$$(-h, \tau), \tau \in (-1, 1), (s(\tau) + h, \tau), \tau \in (0, 1) \; and \; (s(-\tau) + h, \tau), \tau \in (-1, 0), \tag{25}$$

we take the exact solution given by

$$u(x, t) = -1 + exp(t - x + b), (x, t) \in [0, s(t)] \times [0, T = 1] \tag{26}$$

we take initial and boundary conditions

$$u(x, 0) = -1 + exp(b - x), x \in [0, b], s(0) = b \tag{27}$$

$$u(s(t), t) = 0, t \in (0, T = 1], \tag{28}$$

$$\frac{\partial u}{\partial x}(s(t), t) = -s'(t) = -1, t \in (0, T = 1] \tag{29}$$

Random additive noise simulating measurement errors to the Neumann (29) has been added in this example:

$$u_x^\delta(s(t), t) = -1 + N(0, \sigma^2), \tag{30}$$

where $N(0, \sigma^2)$ indicates the normal distribution with mean zero and standard deviation.

$$\sigma = \delta \times max|u_x(s(t), t)| = \delta, t \in (0, 1] \tag{31}$$

where δ is the relative % noise level. Noise could be added in some other quantity related to the position of the moving boundary s(t), but this case is not track here, Fig. 2.

We choose $\lambda = 10^{-6}$ corresponds to the corner of the "L" in Fig. 3, for all three noise levels, we wish to recover the Dirichlet and Neumann boundary conditions in the fixed boundary x = 0 given by

$$u(0, t) = -1 + exp(t + b), t \in (0, 1] \tag{32}$$

$$\frac{\partial u}{\partial x}(0, t) = -1 - exp(t + b), t \in (0, 1] \tag{33}$$

In Figs. 4, 5, 6, and 7 we extend an approximation method of fundamental solutions for $u(0, t)$ and $u_x(0, t)$, respectively, are plotted for two different noise levels $\delta \in 1, 5\%$ and, when compared, they approach the exact solution, however, the accu-

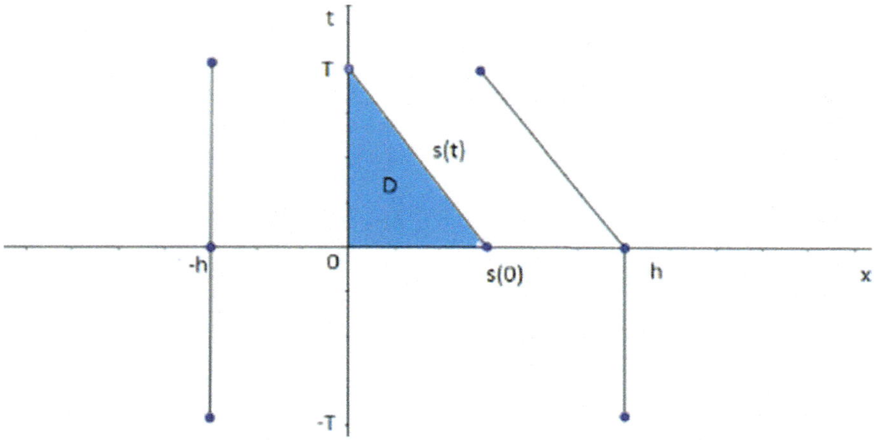

Fig. 2 Particularisation of Fig. 1 for s(t) given by Eq. (24) with b = 0

Fig. 3 L curve plots for $\delta = 1\%$, $\delta = 3\%$, $\delta = 5\%$ when h = 2.5, K = 30 and M = 16

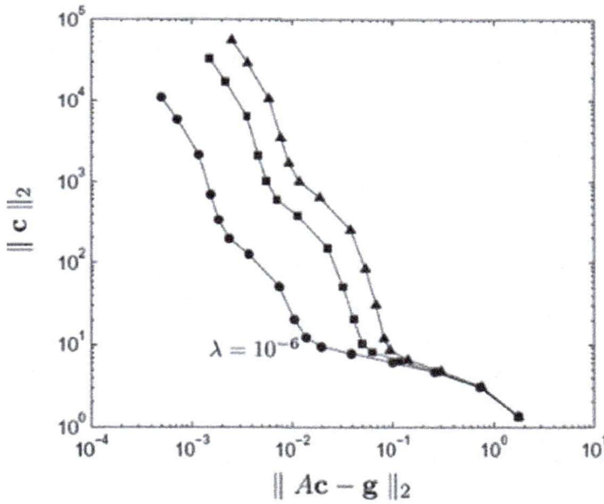

racy deteriorates as time increases, which is predictable, because of the noise and as t increases. These curves also show that, as expected, the heat flux is more difficult to estimate accurately than boundary temperature; however, the numerical solutions are stable and they become more accurate as the amount of noise δ is smaller, Figs. 8, 9, 10, 11 and 12.

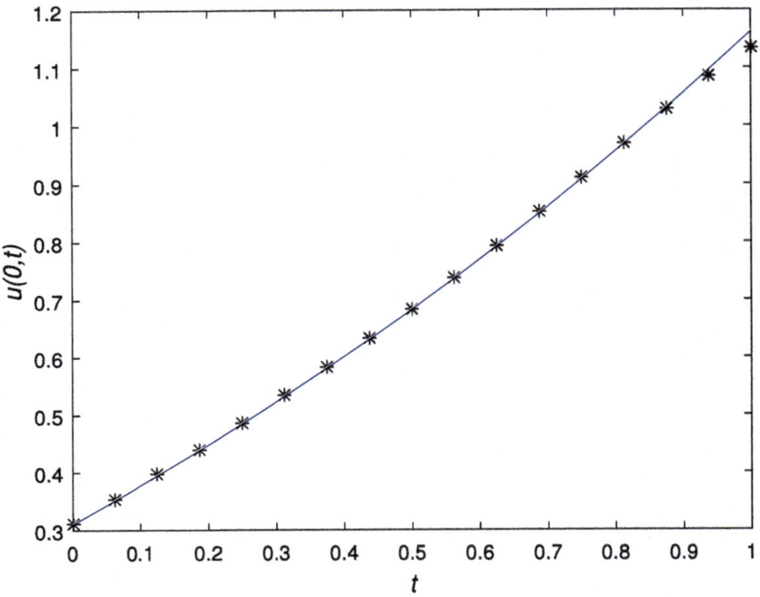

Fig. 4 The exact solution u(0, t) (-) and the MFS approximation. Both plots for $\delta = 0\%$ when h = 2, K = 30 and M = 16

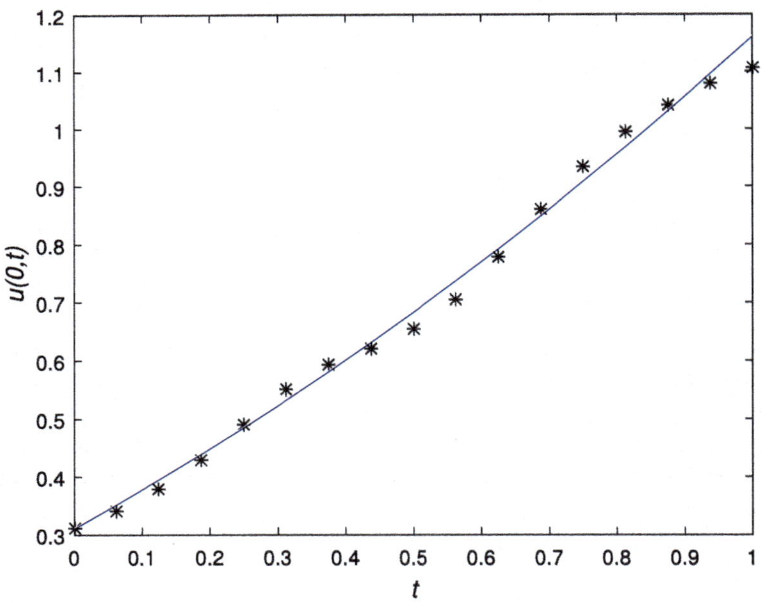

Fig. 5 The exact solution u(0, t) (-) and the MFS approximation. Both plots for $\delta = 5\%$ when h = 2, K = 30 and M = 16

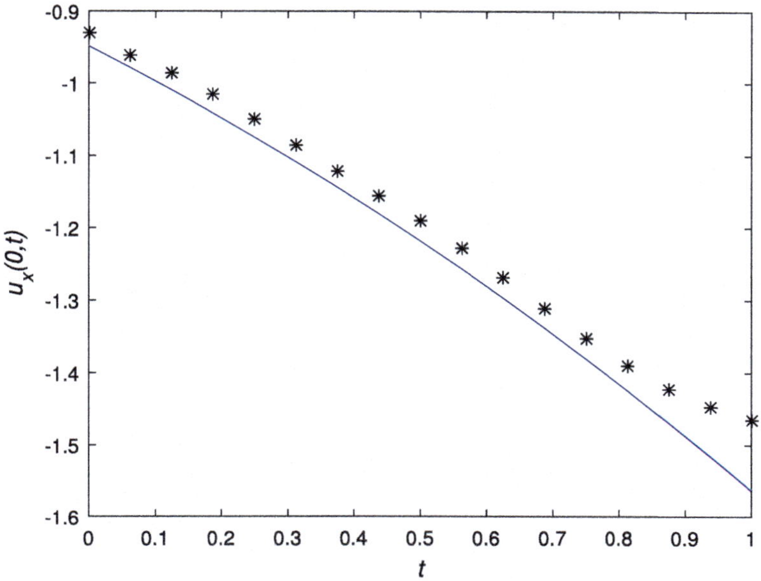

Fig. 6 The exact solution u(0, t) (-) and the MFS approximation. Both plots for $\delta = 0\%$ when h = 2.5, K = 30 and M = 16

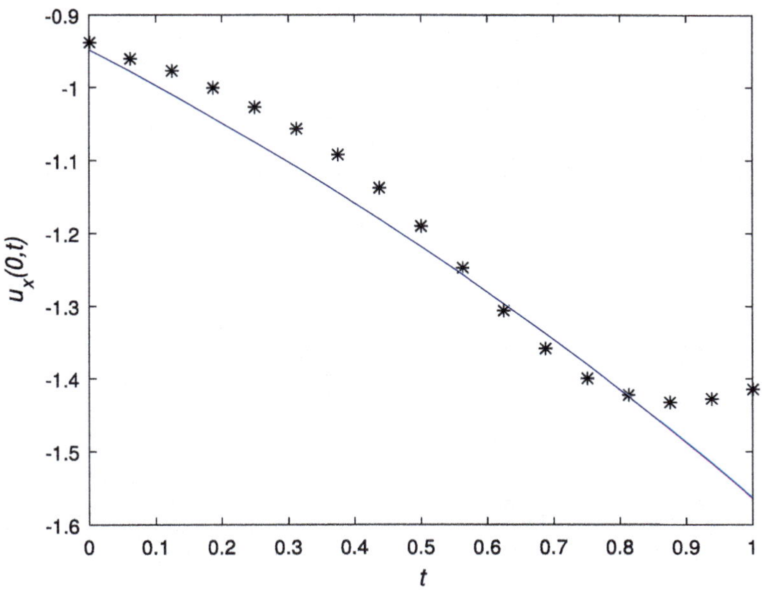

Fig. 7 The exact solution u(0, t) (-) and the MFS approximation. Both plots for $\delta = 5\%$ when h = 2.5, K = 30 and M = 16

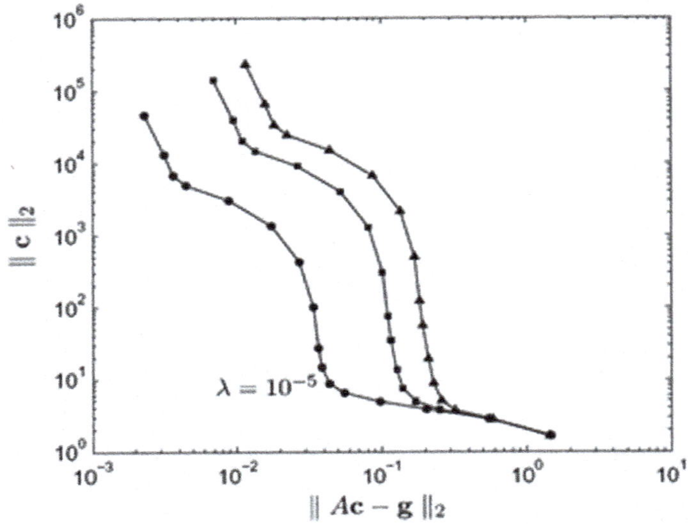

Fig. 8 L curve plots for $\delta = 1\%$, $\delta = 3\%$, $\delta = 5\%$ when h = 2.5, K = 30 and M = 31

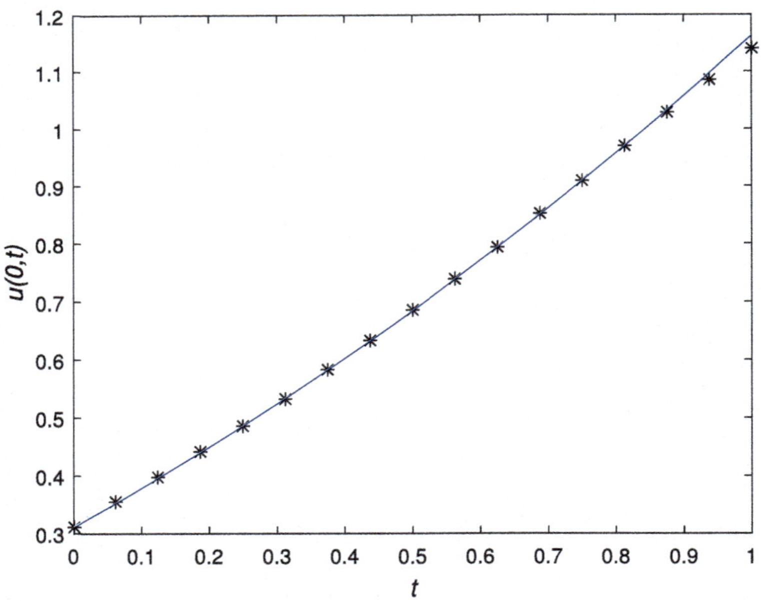

Fig. 9 The exact solution u(0, t) (-) and the MFS approximation. Both plots for $\delta = 0\%$ and obtained with h = 2.5, $\lambda = 10^{-5}$, K = 60 and M = 31

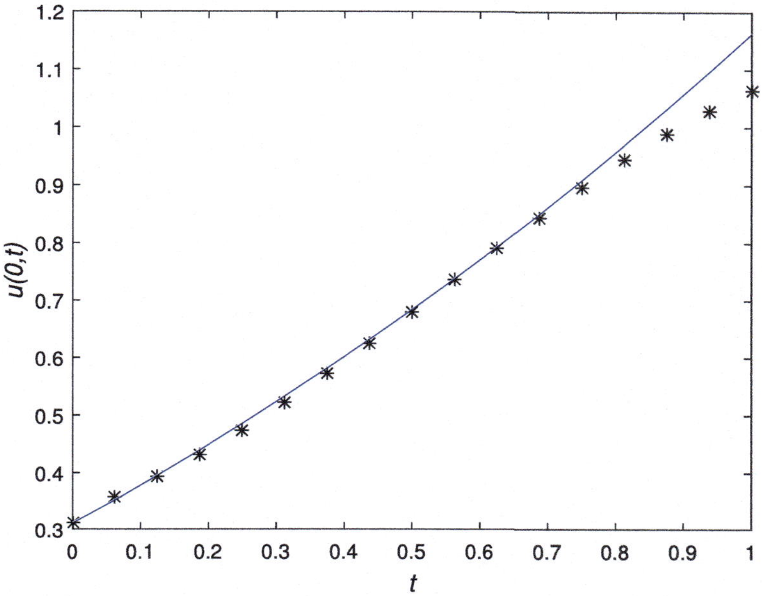

Fig. 10 The exact solution u(0, t) (-) and the MFS approximation. Both plots for $\delta = 5\%$ and obtained with h = 2.5, $\lambda = 10^{-5}$, K = 60 and M = 31

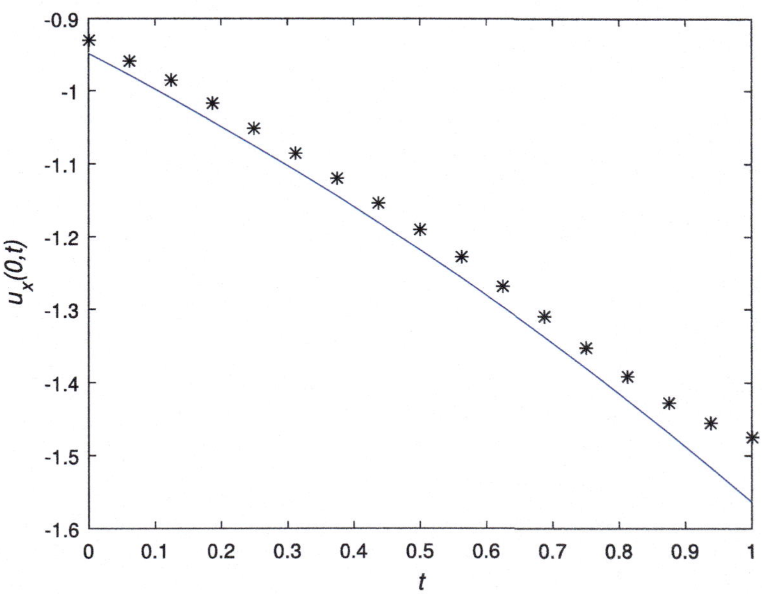

Fig. 11 The exact normal derivative $u_x(0, t)(-)$ and the MFS approximation. Both plots for $\sigma = 0\%$ and obtained with $h = 2.5$, $\lambda = 10^{-5}$, $k = 60$ and $M = 31$

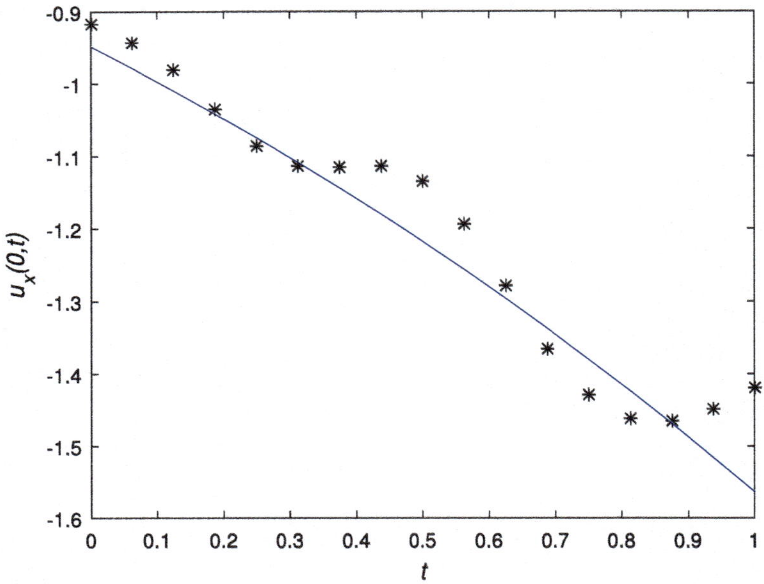

Fig. 12 The exact normal derivative $u_x(0, t)(-)$ and the MFS approximation. Both plots for $\sigma = 5\%$ and obtained with $h = 2.5$, $\lambda = 10^{-5}$, $k = 60$ and $M = 31$

5 Conclusion

In this article, an MFS was proposed and investigated for the one-dimensional inverse Stefan problem, where boundary data are reconstructed with a tikhonov regularization method. Numerical example was presented showing that the MFS to find boundary data can be adjusted to the inverse Stefan problem.

References

1. Rubinstein, L.I.: The Stefan Problem. American Mathematical Society, Providence (1971)
2. Goldman, N.L.: Inverse Stefan Problem. Kluwer Academic Publishers, Dordrecht (1997)
3. Rubinstein, L.: The Stefan problem: comments on its present state. J. Inst. Math. Appl. **24**, 259–277 (1979)
4. Cannon, J.R., Primicerio, M.: Remarks on the one-phase Stefan problem for the heat equation with the flux prescribed on the fixed boundary. Math. Anal. Appl. **35**, 361–373 (1971)
5. Johansson, B.T., Lesnic, D., Reeve, T.: A method of fundamental solutions for the one dimensional inverse Stefan problem. Appl. Math. Modell. **35**, 4367–4378 (2011)
6. Cannon, J.R., van der Hoek, J.: The one phase Stefan problem subject to the specification of energy. J. Math. Anal. Appl. **86**, 281–291 (1982)

7. Cannon, J.R., Hill, C.D.: Existence, uniqueness, stability, and monotone dependence in a Stefan problem for the heat equation. J. Math. Mech. **17**, 1–19 (1967)
8. Cannon, J.R., Primicerio, M.: Remarks on the one-phase Stefan problem for the heat equation with the flux prescribed on the fixed boundary. J. Math. Anal. Appl. **35**, 361–373 (1971)

Comparison Between Gradient Descent and Adam Algorithms for Image Reconstruction in Diffuse Optical Tomography

Nada Chakhim, Mohamed Louzar, Abdellah Lamnii, and Mohammed Alaoui

Abstract In this work, we aim to solve the inverse problem of diffuse optical tomography by using enhanced gradient descent methods. The light propagation throughout the medium is described by the diffusion approximation in frequency domain. For comparison purpose we use the gradient descent method. We have studied the convergence of the objective functional. Our simulation results, in all cases we have tested, show the robustness and the quick convergence of Adam algorithm compared to the gradient descent algorithm.

1 Introduction

Diffuse optical tomography (DOT) is a non- invasive, non-ionized and an inexpensive medical imaging method that uses Near infrared light (NIR) to probe optical properties [1, 2]. The radiative transfer equation (RTE) describes the light propagation [3]. We use the diffusion approximation equation based on the RTE, in the case of frequency domain, throughout this paper.

Early studies used gradient-based methods to solve minimization problems in optical tomography [4–6]. In this work, we adopt Adaptive moment algorithm (Adam) as introduced by Kingma and Lei Ba [7] to recover the optical properties of the DOT inverse problem. Adam is a stochastic gradient based optimization algorithm that only requires first order gradients, and used for problems with sparse or noisy gradients [7, 8].

Our aim is to compare the convergence of the objective functional using Adam and the classical gradient descent (GD) algorithm. We will characterize the performance of the Adam algorithm with respect to the choice of hyper parameters, and in the case

N. Chakhim (✉) · M. Louzar · A. Lamnii · M. Alaoui
Faculty of Science and Technology, University Hassan 1st, Settat, Morocco
e-mail: n.chakhim@uhp.ac.ma

A. Lamnii
LaSAD Laboratory, Ecole Normale Supérieure, Abdelmalek Essaadi University, 93030 Tetouan, Morocco

S. Melliani et al. (eds.), *Applied Mathematics and Modelling in Finance, Marketing and Economics*, Studies in Computational Intelligence 1114,
https://doi.org/10.1007/978-3-031-42847-0_13

of noisy data. A comparison between Adam and gradient descent algorithm will be investigated with respect to the performance measured by the speed of convergence of the objective functional.

This paper is organized as follows: In Sect. 2 we present the mathematical formulation the RTE and its diffusion approximation equation in frequency domain. In Sect. 3 we describe the algorithms used for the recovery of the optical properties of DOT. In Sect. 4 we show the results of our simulation. Finally, we summarize by conclusions.

2 Formulation of the Forward Problem

In this section we firstly describe the mathematical formulation of the RTE based on diffuse optical tomography. Let the space $\mathbb{X}$ be defined by

$$\mathbb{X} := \Omega \times \mathbb{S}^{d-1}$$

where $\mathbb{S}^{d-1}$, $d = 2, 3$ denotes the unit sphere of $\mathbb{R}$, and Ω is the medium of interest which is assumed to be a compact, convex subset of $\mathbb{R}$ with boundary $\partial\mathbb{X} = \partial\mathbb{X}_+ \cup \partial\mathbb{X}_-$

$$\partial\mathbb{X}_\pm := \{(r, \theta) \in \partial\mathbb{X} | \pm (\theta.n) > 0\}$$

where $\partial\mathbb{X}_\pm$ the outgoing and incoming boundaries and n the outward unit normal vector. Photon migration in biological tissues can be described as follows

$$\frac{i\omega}{c}\Phi(r,\theta) + \theta.\nabla\Phi(r,\theta) + (\mu_a + \mu_s)\Phi(r,\theta) = \mu_s \int_{\mathbb{S}^{d-1}} \eta(\theta.\theta')\Phi(r,\theta')\, d\sigma(\theta') + q(r,\theta) \tag{1}$$

where c is the speed of light in medium, i is the imaginary unit, and ω is the angular modulation frequency. The variables r and θ denote the spatial position and the angular direction, respectively. Φ is the radiance, the coefficients μ_a and μ_s are the absorption and scattering coefficients, respectively, and $q(r, \theta)$ is the internal source inside Ω. In this paper,we consider the case with no light source inside $\mathbb{X}$; $q(r, \theta) = 0$

Let $\epsilon_j \subset \partial\Omega$ be the source position, $1 \le j \le s$, s is the number of sources. Then, the boundary condition can be written as

$$\Phi(r,\theta) = \begin{cases} \Phi_0(r,\theta) & \textbf{if} \quad r \in \cup_{j=1}^{s} \epsilon_j \\ \\ 0 & \textbf{if} \quad r \in \partial\Omega \backslash \cup_{j=1}^{s} \epsilon_j \end{cases} \tag{2}$$

This boundary condition implies that if a photon escapes the medium Ω, it does not reenter it. The non-negative normalized phase function $\eta(r, \theta.\theta')$ is the probability that photons traveling in the direction θ' are scattered into the direction θ'.

$$\int_{\mathbb{S}^{d-1}} \eta(\theta.\theta')\, d\sigma(\theta') = \int_{\mathbb{S}^{d-1}} \eta(\theta.\theta')\, d\sigma(\theta) = 1 \tag{3}$$

In optical tomography, the phase function is usually taken as the Henyey-Greenstein phase function [9]. In 2D, it is of the form

$$\eta(\theta.\theta') = \frac{1-g^2}{4(1+g^2-2g\theta.\theta')'} \tag{4}$$

where the parameter $g \in (-1, 1)$ is the scattering shape parameter that describes the shape of the probability density.

The simplest approximation of the RTE described in (1)–(2) is the diffusion approximation, where $\Psi(r) = \int_{\mathbb{S}^{d-1}} \Phi(r, \theta)\, d\sigma(\theta)$. In frequency domain case, The diffusion approximation is of the form

$$-\nabla[.D(r)\nabla\Psi(r)] + (\frac{i\omega}{c} + \mu_a(r))\Psi(r) = q_0(r) \qquad r \in \Omega \tag{5}$$

with the Robin-boundary condition

$$\Psi(r) + 2aD(r)\frac{\partial\Psi(r)}{\partial n} = q(r) \qquad r \in \partial\Omega \tag{6}$$

where a represents the boundary reflection coefficient and depends on the mismatch between the refractive indices, and $D(r)$ is the reduced transport coefficient defined by $D(r) = \frac{1}{3(\mu_a+\mu_s')}$. μ_s' denotes the reduced scattering coefficient expressed as $\mu_s' = (1-g)\mu_s$. We solve The forward model (5)–(6) by using the finite element method for more details we refer the reader to [10].

3 Inverse Problem

Inverse problem we are interested in consists of determining the couple (μ_a, μ_s) from the set of true data y_i such that

$$F_i(\mu_a, \mu_s) = y_i \qquad 1 \le i \le s \tag{7}$$

we denote by F_i the forward operator which is assumed to be Fréchet differentiable, and y_i the approximate measured data. The objective functional can be stated as

$$J(\mu_a, \mu_s) = \frac{1}{2}\sum_{i=1}^{s}(F_i(\mu_a, \mu_s) - y_i)^2 \tag{8}$$

Since the inverse problem is ill-posed, it requires regularization. A first order Tikhonov regularization is applied [11]. By adding a regularization term, the objective functional is formulated as

$$J(\mu_a, \mu_s) = \frac{1}{2}\sum_{i=1}^{s}(F_i(\mu_a, \mu_s) - y_i)^2 + \lambda R(\mu_a, \mu_s) \tag{9}$$

where $R(\mu_a, \mu_s$ is the regularization operator that enforces smoothness conditions in the solution, and λ is the regularization parameter.

The gradient of the objective functional can be written as follows

$$\nabla J(\mu_a, \mu_s) = \sum_{i=1}^{s} F_i'(\mu_a, \mu_s)^*(F_i(\mu_a, \mu_s) - y_i) + \lambda R'(\mu_a, \mu_s) \tag{10}$$

where $R'(\mu_a, \mu_s)$ is the Fréchet derivative of regularisation operator with respect to μ_a. To solve the minimization problem (9) we use Adam algorithm and we compare it with the GD algorithm.

Adam optimizer

Adam method is given by

$$x^{k+1} = x^k - \beta\frac{\hat{m_{k+1}}}{\sqrt{\hat{v_{k+1}}} + \epsilon} \tag{11}$$

where

$$\hat{m_k} = \frac{m_k}{(1-\rho_1^k)} \tag{12}$$

$$\hat{v_k} = \frac{v_k}{(1-\rho_2^k)} \tag{13}$$

$$m_k = \rho_1.m_{k-1} + (1-\rho_2).g_k \tag{14}$$

$$v_k = \rho_1.v_{k-1} + (1-\rho_2).g_k^2 \tag{15}$$

$$g_k = \nabla J \tag{16}$$

with β is the step size parameter, ρ_1 and ρ_2 are the exponential decay rates for the moment estimates.

Gradient descent optimizer

The gradient descent method can be described as follow

$$x^{k+1} = x^k - \tau\nabla J \tag{17}$$

with τ is the step size parameter.

Convergence of Adam algorithm in DOT problem

The convergence of Adam algorithm is well established in [12] under the assumption that the objective function has a bounded gradient. Here we show the compatibility of this algorithm with the problem of DOT.

The forward operator F_i and all its Fréchet derivatives are continuous, then, $(F_i - y_i)$ is bounded. Moreover, the adjoint Fréchet derivative F_i' is continuous and bounded [11]. Thus, the gradient of the objective functional $J(\mu_a)$ is bounded, then the convergence of Adam algorithm for the DOT problem is obtained.

4 Numerical Results

The synthetic data is generated by using the Toast++ software [13], which solve the forward problem (5)–(6) described in previous section. In all simulation cases, a circular domain which contains different inclusion sizes is performed. 20 sources and 20 detectors are equally spaced on the boundary of the domain. the frequency parameter ω is equal to 100 Mhz. The regularization parameter λ is set to be equal to 10^{-8}. The background medium has absorption coefficient of $\mu_a = 0.015\,\text{mm}^{-1}$ and reduced scattering coefficient of $\mu_s' = 1\,\text{mm}^{-1}$, respectively, and these values keep the same throughout this paper. The initial guess of reconstruction is set to be identical to the properties of the background medium. For the choice of the hyper parameters of the Adam algorithm in all cases, we take $\rho_1 = 0.85$, $\rho_2 = 0.95$ and $\epsilon = 10^{-8}$. The parameter β is chosen empirically in the range of 0.001 to 1 such that we obtain $\frac{\|F_i(\mu_a^{true}) - F_i(\mu_a^{recon})\|^2}{\|F_i(\mu_a^{true})\|^2} \leq \delta$ with less iteration count.

Figure 1 shows the results of one inclusion. Figure 1a shows the true distribution of the absorption coefficient. Figure 1b shows the reconstruction of the absorption coefficient using Adam algorithm after 63 iterations with $\beta = 0.1$. Figure 1c shows the reconstruction of the absorption coefficient using GD algorithm after 400 iterations with $\tau = 0.9$. By comparing Fig. 1b and c, we observe that the reconstruction of the shape of the absorption coefficient seems to be satisfactory. The internal values are closer to the true distribution values.

Figure 2 shows the case of two inclusions with different sizes. Figure 2a shows the true distribution of the absorption coefficient. Figure 2b shows the reconstruction of the absorption coefficient using Adam algorithm after 80 iterations with $\beta = 0.15$. Figure 2c shows the reconstruction of the absorption coefficient using GD algorithm after 500 iterations with $\tau = 1.5$.

In Fig. 3 we have contaminated the data with 0.1% additive random Gaussian noise. Figure 3a shows the true distribution of the absorption coefficient. Figure 3b shows the reconstruction using Adam algorithm with $\beta = 0.28$. Figure 3c shows the reconstruction using GD algorithm with $\tau = 3$. We observe that the quality of image is low when noise is added.

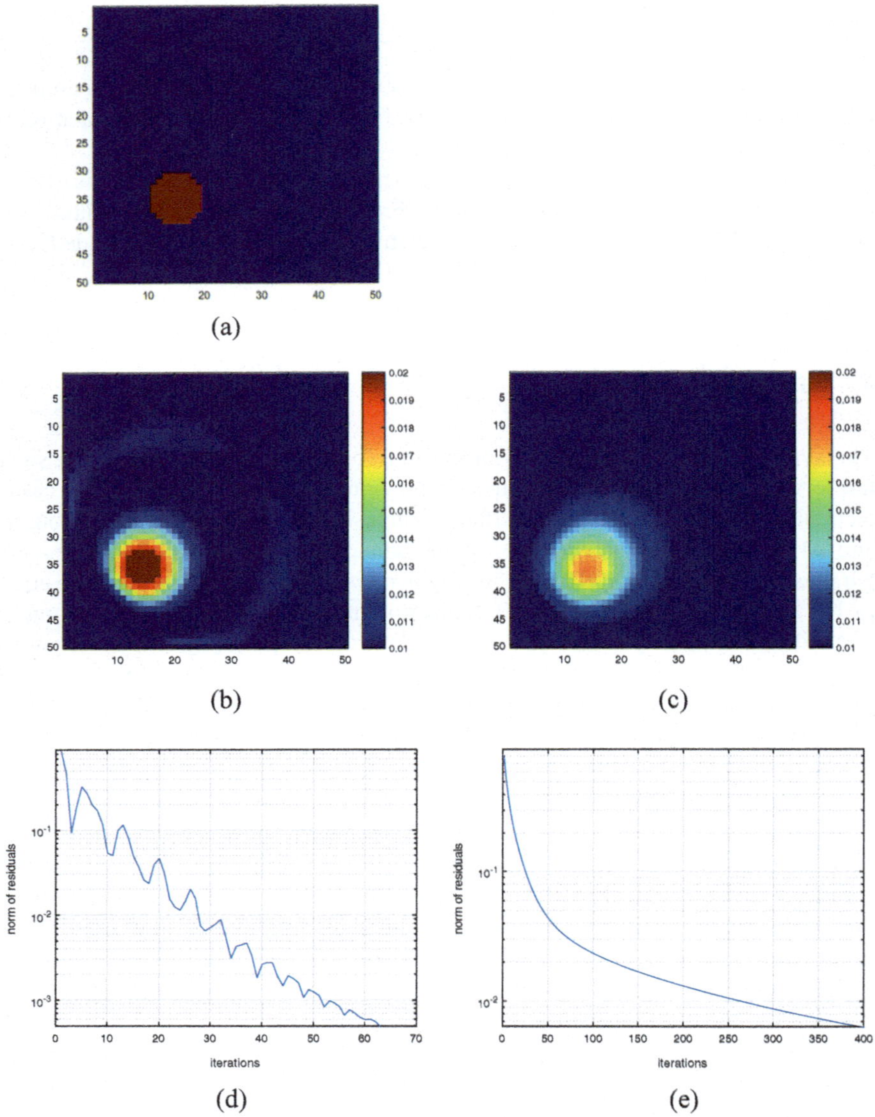

Fig. 1 Reconstructed images of μ_a for the case of one inclusion. **a** True distribution of μ_a. **b** Reconstruction using Adam algorithm. **c** Reconstruction using GD algorithm. **d** Norm of residuals of Objective functional using the Adam algorithm. **e** Norm of residuals of Objective functional using the GD algorithm

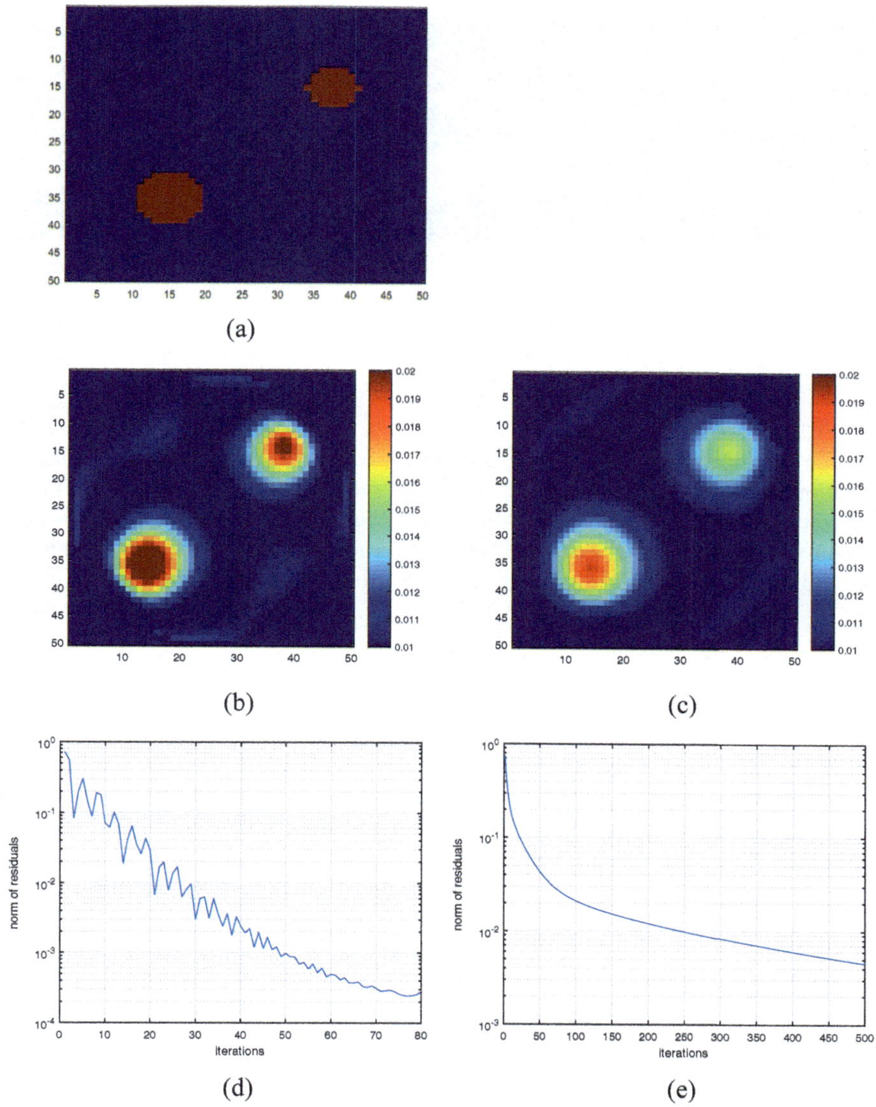

Fig. 2 Reconstructed images of μ_a for the case of two inclusions. **a** True distribution of μ_a. **b** Reconstruction using Adam algorithm. **c** reconstruction using GD algorithm. **d** Norm of residuals of Objective functional using the Adam algorithm. **e** Norm of residuals of Objective functional using the GD algorithm

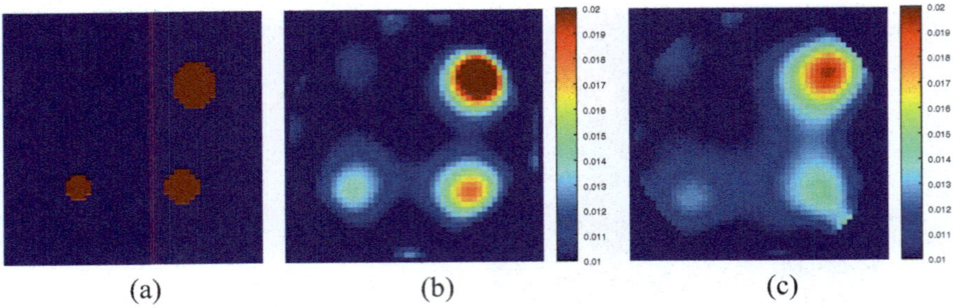

Fig. 3 Reconstructed images of μ_a for the case of three inclusions using Adam algorithm by adding 0.1% random Gaussian noise. **a** True distribution of μ_a. **b** Reconstruction using Adam algorithm. **c** Reconstruction using GD algorithm

The reconstruction results obtained prove that Adam algorithm can localize the position of inclusions for different sizes and the algorithm converges quickly as expected. However, the GD algorithm seems to be less efficient and computationally expensive, especially, in the case of small inclusions. We have noticed that the number of iterations depends on the number and size of inclusions.

5 Conclusion

In this study, we have applied two optimizers algorithms to recover the shape of absorption coefficient in diffuse optical tomography. In the first step, we tested this algorithms on a free noise synthetic data and then we added some Gaussian noise. The numerical simulation results show that Adam algorithm can successfully reconstruct the shape and locations of inclusions for all cases we have simulated, and outperforms the gradient descent algorithm. It is observed that Adam algorithm is less sensitive to noise.

Conflicts of Interest The authors declare that they have no conflicts of interest regarding the publication of this paper.

References

1. Gopinath, S., Robertson, C.S., Grossman, R.G., Chance, B.: Near-infrared spectroscopic localization of intracranial hematomas. J. Neurosurg. **79**, 43–47 (1993)
2. Taroni, P., et al.: Noninvasive assessment of breast cancer risk using time-resolved diffuse optical spectroscopy. J. Biomed. Opt. **15**(6), 060501 (2010)
3. Klose, A.D., Hielscher: Iterative reconstruction scheme for optical tomography based on the equation of radiative transfer. Med. Phys. (1999)
4. Arridge, S.R., Schotland, J.C.: Optical tomography: forward and inverse problems. In: Inverse Problems (2009)

5. Roy, R., Sevick-Muraca, E.M.: A numerical study of gradient based nonlinear optimization methods for contrast enhanced optical tomography. Opt. Express **9**(1), 49–65 (2001)
6. Arridge, S.R.: Optical tomography in medical imaging. Inverse Probl. **15**(2), R41–R93 (1999)
7. Kingma, D.P., Lei Ba, J.: Adam: a method for stochastic optimization. In: Conference Paper at ICLR (2015)
8. Ruder, S.: An overview of gradient descent optimization algorithms (2017)
9. Henyey, L.C., Greenstein, J.L.: Diffuse radiation in the galaxy. Astrophys. J. **93**, 70–83 (1941)
10. Arridge, S.R., Schweiger, M., Hiraoka, M., Delpy, D.T.: A finite element approach for modeling photon transport in tissue. Med. Phys. **20**(2), 299–309 (1993)
11. Scherzer: Convergence rates of iterated Tikhonov regularized solutions of nonlinear Ill - posed problems. Numerische Mathematik (1993)
12. Reddi, S.J., Kale, S., Kumar, S.: On the convergence of adam and beyond. In: Conference Paper at ICLR (2018)
13. Schweiger, M., Arridge, S.R.: The Toast++ software suite for forward and inverse modeling in optical tomography. J. Biomed. Opt. **19**(4), 040801 (2014)

Modelling and Forecasting Individuals Using the Internet (% of Population) in Morocco

Oussama Rida, Ahmed Nafidi, and Boujemaa Achchab

Abstract This study introduces a new stochastic diffusion process that is based on the generalized Goel-Okumoto curve. By analyzing the corresponding stochastic differential equation (SDE), we can accurately determine the probabilistic characteristics of the process, including its solution, transition density probability function, and distribution. To estimate the model parameters, we employ the maximum likelihood method and utilize discrete sampling. This allows us to formulate a nonlinear equation that can be efficiently solved using metaheuristic optimization algorithms like simulated annealing. The proposed model is then applied to fit and forecast data on Individuals using the Internet (% of population) in Morocco.

1 Introduction

Following the "privatization" of the Internet in the United States during the mid-1990s, there has been a significant expansion of the interconnected network. This expansion has paved the way for a wave of innovation in information technology, as well as in various fields that make use of this technology. As a result, a wide range of online services and new "business models" have emerged, transforming the way we interact and conduct business in today's digital age. During the same period, the United States experienced remarkable economic growth without sig-

O. Rida (✉) · A. Nafidi · B. Achchab
National School of Applied Sciences, LAMSAD, Hassan First University of Settat, Berrechid, Morocco
e-mail: oussamar@gmail.com

A. Nafidi
e-mail: ahmed.nafidi@uhp.ac.ma

B. Achchab
e-mail: boujemaa.achchab@uhp.ac.ma

B. Achchab
Mohammed VI Polytechnic University, Ben Guerir, Morocco

S. Melliani et al. (eds.), *Applied Mathematics and Modelling in Finance, Marketing and Economics*, Studies in Computational Intelligence 1114,
https://doi.org/10.1007/978-3-031-42847-0_14

nificant inflationary pressures. This coincided with the emergence of the Internet as a central component of a new growth paradigm, often referred to as the "new economy". This perception contributed to the formation and subsequent rapid expansion of a speculative bubble that heavily impacted internet-based businesses, [2].

In his work, David [4, 5] characterizes digital technology as a general-purpose technology that has a significant impact on economic performance. The influence of digital technology extends to various aspects of the economy and society, including consumption patterns, production methods, and organizational structures. Similar to the transformative inventions of the late nineteenth century, digital technology is expected to bring about fundamental changes in the economy, stimulate economic growth, and reshape society. However, these effects are anticipated to materialize over the long term rather than immediately.

The growth phenomena being investigated exhibit two distinctive characteristics. Firstly, they are dynamic rather than static, meaning that they involve changes over time. Secondly, they are complex and involve multiple variables, some of which may be unknown or difficult to quantify. To address these challenges, stochastic processes, including diffusion processes, can be utilized. In such processes, random fluctuations are incorporated into the differential equations, representing the curves associated with the growth phenomena. The solution to these equations provides the analyzed curve, thus incorporating stochastic differential equations into the modeling framework.

This approach has generated a wide variety of works relating to the growth curves of which we can only list some. Regarding the Gompertz curve, a diffusion process related to the curve was the first to be considered in Capocelli and Ricciardi [3]. This has been extended to other areas of study such as population growth. Another applications, such as demographics (Artzrouni and Reneke) [1] and Age dependency ratio (% of working-age population) in Morocco (Nafidi et al.) [13] or energy consumption (Giovanis and Skiadas) [6], are linked to some variants suggested by Tuckwell and Koziol [18].

The process examined in this work is the stochastic diffusion process based on generalized Goel-Okumoto curve which is in software reliability engineering used for example, by Goel and Okumoto [8], which has been further generalized in Goel [7] and modified by many researchers in developing a model of the classification theme according to the nature of the debugging strategy. Thus, the generalized Goel-Okumoto curve growth curve is given by :

$$y(t) = b_0(1 - e^{-bt})^p \tag{1}$$

The mean curve of the process represents the central tendency of the variable under study, and its range can be influenced by the initial value. These dynamic models excel in their ability to make predictions about future behavior. However, to effectively apply this model, precise parameter estimates are required for accurate implementation. In our case, for addressing the prediction issue, we employ the maximum likelihood method. While explicit expressions are available for calculating the parameters of the initial process distribution, the same approach cannot be applied to the remaining parameters. For those, a complex system of equations arises, and finding a solution through classical numerical methods is not guaranteed. To overcome this challenge, we propose the utilization of metaheuristic optimization algorithms such as simulated annealing (SA).

This section is organized as follows: firstly, we define the stochastic diffusion process based on the generalized Goel-Okumoto curve as a solution to Ito's stochastic differential equation (SDE). Next, by applying Ito's formula, we derive the analytical expression of this process. Subsequently, we determine the trend and conditional trend functions associated with the process. In Sect. 3, we focus on estimating the parameters of the proposed process through optimization methods. Specifically, we employ simulated annealing as a local search method within the variable neighborhood search framework to estimate the parameters based on the log-likelihood equation. This allows us to obtain confidence intervals for the predicted values. In the final section, we apply the model to real-time data of Individuals using the Internet (% of population) in Morocco. The results demonstrate satisfactory forecasting accuracy, indicating the effectiveness of the proposed model.

2 The Model and Its Probabilistic Characteristics

The model is a diffusion process that operates in one dimension and has a range of values between 0 and infinity. It is defined by the process $x(t)$ for time intervals within $[t_0, T]$, which satisfies a nonlinear stochastic differential equation (SDE) given as:

$$dx(t) = \frac{pb}{e^{bt} - 1} x(t)dt + \sigma x(t)dw(t), \quad x(t_0) = x_{t_0}, \tag{2}$$

Here, the parameters b, p, and σ are positive real numbers, and $w(t)$ represents a one-dimensional standard Wiener process.

2.1 Analytical Representation of the Model

We elucidate the primary attributes of the process, specifically in the context of prediction. These attributes encompass the mean function, which, due to the model's configuration, adopts a Generalized Goel Okumoto curve, making it particularly well-suited for both fitting and forecasting.

By transforming the SDE Eq. (2) into $y(t) = \log(x(t))$ and applying the Itô formula, we obtain the analytical expression.

$$y(t) = y_0 + p \log\left(\frac{1 - e^{-bt}}{1 - e^{-bt_0}}\right) - \frac{\sigma^2}{2}(t - t_0) + \sigma(w(t) - w(t_0))$$

By substituting, we have the following expression:

$$x(t) = x_0 \left(\frac{1 - e^{-bt}}{1 - e^{-bt_0}} \right)^p \exp \left(-\frac{\sigma^2}{2}(t - t_0) + \sigma(w(t) - w(t_0)) \right), t \geq 0 \quad (3)$$

2.2 *Distribution of the Process*

For $s < t$, the variable $y(t) \mid y(s) = y_s$ follows a normal distribution with mean $g(s, t, x_s)$ and variance $\sigma^2(t - s)$. Therefore, the $x(t) \mid x(s) = x_s$ is given by $\Lambda[g(s, t, x_s), \sigma^2(t - s)]$, where $g(s, t, x_s) = \log(x_s) + p \log\left(\frac{1-e^{-bt}}{1-e^{-bs}}\right) - \frac{\sigma^2}{2}(t - s)$, see [15].

Then, the transition probability density function (TPDF) of the process is : for $s < t$

$$f(x, t \mid y, s) = \frac{1}{x\sqrt{2\pi\sigma^2(t - s)}} \exp \left(- \frac{\left[\log(\frac{x}{y}) - p \log\left(\frac{1-e^{-bt}}{1-e^{-bs}}\right) + \frac{\sigma^2}{2}(t - s)\right]^2}{2\sigma^2(t - s)} \right) \quad (4)$$

3 Moments of the Process

The function that describes the conditional trend function (CTF) of the process is given by:

$$\mathbb{E}[x(t) \mid x(s) = x_s] = x_s \left(\frac{1 - e^{-bt}}{1 - e^{-bs}} \right)^p, \quad t > s, \quad (5)$$

Furthermore, considering the initial condition $P[x(t_0) = x_0] = 1$, the function that describes the overall trend of the process is known as the trend function (TF).

$$\mathbb{E}[x(t)] = x_0 \left(\frac{1 - e^{-bt}}{1 - e^{-bt_0}} \right)^p, \quad t \geq t_0, \quad (6)$$

- Without the presence of random fluctuations (i.e., when $\sigma = 0$), through a straightforward integration, the solution of the ordinary differential equation (ODE) linked to the stochastic differential equation (SDE) described in Eq. (2) becomes $x(t) = k(1 - e^{-bt})^p$. This solution bears a resemblance to the generalized Goel-Okumoto curve expressed in Eq. (1).

Furthermore, we can calculate the quantile function by using the estimated parameters, where z_α represents the α-quantile of a standard normal distribution.

$$P_\alpha(t) = x_0 \left(\frac{1 - e^{-bt}}{1 - e^{-bt_0}}\right)^p \exp\left\{-\frac{\sigma^2}{2}(t - t_0) + z_\alpha \sigma \sqrt{t - t_0}\right\} \tag{7}$$

4 Estimation of Parameters

Given the explicit expression of the probability density function (TPDF), we can employ the maximum likelihood method to estimate the parameters b, p, and σ^2. This estimation is based on a discrete sampling of the process at specific time instances $t_1, t_2, \ldots, t_n$, denoted as $x_{t_1}, x_{t_2}, \ldots, x_{t_n}$. It is assumed that the time intervals between consecutive samples, $t_i - t_{i-1}$ for $i = 2, \ldots, n$, are constant and denoted as h. In this context, the notation $x_{t_i} = x_i$ is used. Assuming the initial condition $\mathbb{P}[x(t_1) = x_1] = 1$, we can derive the associated likelihood function from Eq. (4).

$$\mathbb{L}(x_1, x_2, \ldots, x_n; b, p, \sigma^2) = \prod_{i=2}^{n} f(x_i, t_i \mid x_{i-1}, t_{i-1}).$$

For computational convenience, we utilize the log-likelihood function to simplify the above function.

$$\begin{aligned}\log \mathbb{L}(x_1, x_2, \ldots, x_n; b, p, \sigma^2) = &-\frac{n-1}{2}\log(2\pi h) - \frac{n-1}{2}\log\sigma^2 - \sum_{i=2}^{n}\log x_i \\ &-\frac{1}{2h\sigma^2}\sum_{i=2}^{n}\left[\log\left(\frac{x_i}{x_{i-1}}\right) - p\log\left(\frac{1 - e^{-bt_i}}{1 - e^{-bt_{i-1}}}\right) + \frac{\sigma^2}{2}h\right]^2.\end{aligned} \tag{8}$$

By taking the derivatives of the log-likelihood function with respect to b, p, and σ^2, with $t_i - t_{i-1}$, and applying the principle of maximum likelihood, the following relationship is obtained:

$$\begin{aligned}\frac{\partial \log(\mathbb{L}(b, p, \sigma^2))}{\partial b} = &\sum_{i=2}^{n} \frac{\log\left(\frac{x_i}{x_{i-1}}\right) - p\log\left(\frac{1-e^{-bt_i}}{1-e^{-bt_{i-1}}}\right) + \frac{\sigma^2}{2}h}{h} \\ &\times \frac{t_{i-1}e^{-bt_{i-1}-1}(1 - e^{-bt_i}) - t_i e^{-bt_i-1}(1 - e^{-bt_{i-1}})}{(1 - e^{-bt_i})(1 - e^{-bt_{i-1}})} = 0\end{aligned} \tag{9}$$

$$\frac{\partial \log(\mathbb{L}(b,p,\sigma^2))}{\partial p} = \sum_{i=2}^{n} \frac{\log\left(\frac{x_i}{x_{i-1}}\right) - p\log\left(\frac{1-e^{-bt_i}}{1-e^{-bt_{i-1}}}\right) + \frac{\sigma^2}{2}h}{h} \log\left(\frac{1-e^{-bt_i}}{1-e^{-bt_{i-1}}}\right) = 0$$

$$\frac{\partial \log(\mathbb{L}(b,p,\sigma^2))}{\partial \sigma^2} = \sigma^4 h^2(n-1) + 4\sigma^2 h(n-1) - 4\sum_{i=2}^{n} \log^2\left(\frac{x_i}{x_{i-1}}\right)$$

$$-4p^2 \sum_{i=2}^{n} \left(\log\left(\frac{1-e^{-bt_i}}{1-e^{-bt_{i-1}}}\right)\right)^2 + 8p\sum_{i=2}^{n} \log(\frac{1-e^{-bt_i}}{1-e^{-bt_{i-1}}}) \log(\frac{x_i}{x_{i-1}}) = 0 \quad (10)$$

We use the discriminant for Eq. (10) and after some calculation the estimator of $\hat{\sigma}$ is:

$$\hat{\sigma}^2 = \frac{2}{h}\left[\left(1 + \frac{1}{n-1}\sum_{i=2}^{n}\left(\log\left(\frac{x_i}{x_{i-1}}\right) - p\log\left(\frac{1-e^{-bt_i}}{1-e^{-bt_{i-1}}}\right)\right)^2\right)^{\frac{1}{2}} - 1\right] \quad (11)$$

For the estimator of p and b, we need numerical methods to approximate the solutions for $\gamma(b)$ and $\phi(p)$

$$\gamma(b) = \sum_{i=2}^{n}\left[\log\left(\frac{x_i}{x_{i-1}}\right) - p\log\left(\frac{1-e^{-bt_i}}{1-e^{-bt_{i-1}}}\right) + \frac{\hat{\sigma}^2}{2}h\right]\left[\frac{t_{i-1}e^{-bt_{i-1}}}{1-e^{-bt_{i-1}}} - \frac{t_i e^{-bt_i}}{1-e^{-bt_i}}\right] \quad (12)$$

$$\phi(p) = \sum_{i=2}^{n}\left[\log\left(\frac{x_i}{x_{i-1}}\right) - p\log\left(\frac{1-e^{-bt_i}}{1-e^{-bt_{i-1}}}\right) + \frac{\hat{\sigma}^2}{2}h\right]\left[\log\left(\frac{1-e^{-bt_i}}{1-e^{-bt_{i-1}}}\right)\right] \quad (13)$$

In our specific case, we will estimate the parameters using Simulated Annealing (SA) applied to the log-likelihood equation provided below:

$$T(b,p,\sigma^2) = -\frac{n-1}{2}\log\sigma^2 - \frac{1}{2h\sigma^2}\sum_{i=2}^{n}\left[\log\left(\frac{x_i}{x_{i-1}}\right) - p\log\left(\frac{1-e^{-bt_i}}{1-e^{-bt_{i-1}}}\right) + \frac{\sigma^2}{2}h\right]^2 \quad (14)$$

4.1 Aspects on Optimization Methods

As stated above, the use of stochastic optimization methods will be an alternative to solving log-likelihood equation, such as Simulated Annealing (SA). The proposed algorithm is designed to address optimization problems of the form min $f(w)$, where $w \in \Omega$. In certain scenarios, this algorithm is recommended over traditional numerical methods as it imposes fewer constraints on the solution space and analytical properties of the objective function. In the context of maximum likelihood estimation for probability distributions, similar approaches have been utilized in previous studies. For example, Nafidi et al. [15], Vera and Dìaz-Garcìa [19] as well as Román-Romá, Torres-Ruiz, and Francisco [16] have employed comparable methods in their respective investigations.

4.1.1 Simulated Annealing (SA) Algorithm

The algorithm under consideration is a local search metaheuristic introduced by Kirkpatrick [11], inspired by the annealing process observed in materials science and mechanical statistics. In a general sense, the algorithm follows the following procedure:

Given a solution θ at a particular iteration and the corresponding value of the objective function $f(\theta)$, in the subsequent iteration, a new value θ' is selected from the neighborhood N_θ of θ. The increase in the objective function, denoted as $\Delta = f(\theta') - f(\theta)$, is evaluated. If $\Delta \leq 0$, θ' is chosen as the new solution. Otherwise, there is a probability $p = \exp(-\Delta/T)$, where T represents the temperature, that θ' is accepted. Consequently, an internal loop generates a Markov chain, the length of which corresponds to the number of loop iterations. At the conclusion of each loop, the temperature gradually decreases, leading to the generation of a new Markov chain. Initially, during the cooling process, solutions that result in a deterioration of the objective function are allowed (at higher temperatures). However, as the temperature decreases, such solutions are increasingly less accepted.

4.2 Estimated TF and Estimated CTF

Based on Zenha's theorem [20], we can derive the estimated conditional trend (ECTF) by substituting the parameter estimators into Eqs. (5) and (6).

$$\hat{E}[x(t)|x(s) = x_s] = x_s \left(\frac{1 - e^{-\hat{b}t}}{1 - e^{-\hat{b}s}} \right)^{\hat{p}} \tag{15}$$

and the estimated trend function (ETF) is given by:

$$\hat{E}[x(t)] = x_{t_0} \left(\frac{1 - e^{-\hat{b}t}}{1 - e^{-\hat{b}t_0}} \right)^{\hat{p}} \tag{16}$$

Moreover, the estimated quantile is derived by substituting the parameter estimates into Eq. (7).

$$\hat{P}_\alpha(t) = x_0 \left(\frac{1 - e^{-\hat{b}t}}{1 - e^{-\hat{b}t_0}} \right)^{\hat{p}} \exp\left\{ -\frac{\hat{\sigma}^2}{2}(t - t_0) + z_\alpha \hat{\sigma} \sqrt{t - t_0} \right\} \tag{17}$$

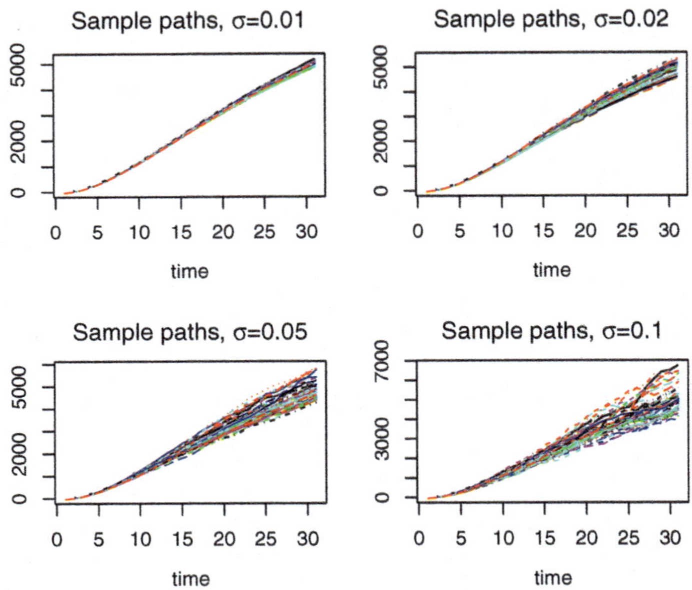

Fig. 1 Simulated sample path of Generalized Goel-Okumoto process

5 Simulation

We employ the algorithm based on the numerical solution of the SDE associated with the process, as described in [12]. We simulate 50 trajectories of the process, governed by Eq. (3), where each trajectory consists of 30 observations within the interval [0.05, 10].

Figure 1 illustrates the process with the following initial values: $x_0 = 5, t_0 = 0.05$, $b = 0.5$, and $p = 2$. The plot includes multiple curves for different values of σ.

6 Application to Real-World Data

The proposed model is implemented to analyze the data related to the percentage of individuals using the Internet in Morocco from the years 2002 to 2018. Internet users are individuals who have accessed and utilized the Internet within the past three months. This includes various devices such as computers, mobile phones, personal digital assistants, gaming consoles, digital televisions, and other compatible devices.

In the past, telecommunication operators have emerged as the primary data source for telecommunications information. This has facilitated easy access to data on subscriptions for a majority of countries. Although it provides a broad overview of accessibility, a more precise indicator is the penetration rate, which represents the proportion of households that have access to telecommunications services. This metric offers a more accurate assessment of the extent to which households are equipped with and utilizing telecommunications facilities. Over the past few years, there has been an increase in the availability of research and communication technology data obtained through surveys conducted with households and businesses. Equally significant are the data concerning the actual utilization of telecommunications services. Ideally, statistics encompassing all three dimensions, namely subscriptions, access, and usage, should be compiled for comprehensive insights into the telecommunications sector. It should be noted that the quality of data may vary among countries due to discrepancies in regulations pertaining to data provision and accessibility. Inconsistencies can occur when the fiscal year of a country does not align with the calendar year used by the International Telecommunication Union (ITU), which concludes at the end of December each year. Several countries have fiscal years that conclude in March or June, creating potential discrepancies in reporting and data collection [17].

The Internet is a globally accessible computer network that provides access to various communication services, such as the World Wide Web. It facilitates the transmission of email, news, entertainment, and data files, regardless of the device used (not limited to computers but also including mobile phones, PDAs, gaming machines, digital TVs, etc.). Access to the Internet can be achieved through either a fixed or mobile network. For the most recent and comprehensive information on sources and country-specific details. For additional/latest information on sources and country notes.[1]

The data, shown in Table 1, was obtained from the World Bank's database.[2] The method used consists of two phases:

- The data from 2002 to 2016 is utilized to estimate the parameters of the process using simulated annealing and the R programming language. The resulting parameter estimates are as follows: $\hat{b} = 1.7012$, $\hat{p} = 1.75$, and $\hat{\sigma} = 0.032$.
- The data spanning from 2017 to 2018 is examined in order to predict the anticipated values of the process. The results, summarized in Table 1, provide an overview of the behavior of the conditional and non-conditional trend functions, as well as the quantile values presented in Table 2, calculated using Eqs. (16), (15), and (17). The performance of the short-term forecast for the process is depicted in Figs. 2 and 3.

In all scenarios, the values fitted with forecasts based on the ETF exhibit slightly better accuracy compared to the ECTF. To assess the model's performance, we calculate two error metrics: the Mean Absolute Percentage Error (MAPE) and the Symmetric Mean Absolute Percentage Error (SMAPE). These metrics are defined as follows:

---

[1] https://itu.int/en/ITU-D/Statistics/Pages/stat/default.aspx. Infrastructure: Communications, 2020.

[2] https://data.worldbank.org. World bank.individuals using the internet (% of population) - morocco, 2019.

Table 1 Individuals using the Internet (% of population), ETF and ECTF

Years	$x(t)$	ETF	ETCF
2002	2.37	2.37	2.37
2003	3.35	2.81	2.81
2004	11.6	8.2	9.7
2005	15.08	14.52	20.55
2006	19.77	21.01	21.82
2007	21.5	27.27	25.65
2008	33.1	33.08	26.08
2009	41.3	38.35	38.37
2010	52	43.06	46.36
2011	46.1	47.21	57.01
2012	55.41	50.84	49.65
2013	56	53.99	58.84
2014	56.8	56.71	58.82
2015	57.08	59.05	59.14
2016	58.27	61.05	59.01
Forcast			
2017	61.76	62.76	59.90
2018	64.8	64.23	61.30

Table 2 Forcasted ETF and Inferior limit - Upper limit

Years	ETF	Quantile
2017	62.76	(45.03–97.67)
2018	64.23	(45.24–95.03)

$$MAPE = \frac{1}{N}\sum_{i=1}^{N}\frac{\left|x(t_i) - \hat{x}(t_i)\right|}{x(t_i)} * 100$$

$$SMAPE = \frac{1}{N}\sum_{i=1}^{N}\frac{\left|x(t_i) - \hat{x}(t_i)\right|}{(|x(t_i)| + |\hat{x}(t_i)|)/2} * 100$$

The accuracy of the forecast can be evaluated based on the MAPE and SMAPE results, which are 7% and 8%, respectively. Both values are below the threshold of 10%, indicating that the forecast is highly accurate. This assessment aligns with the findings presented in [9].

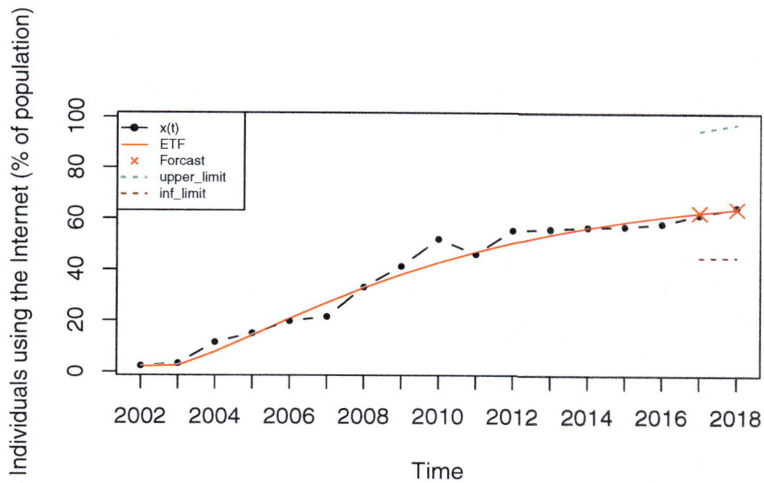

Fig. 2 Real data and estimated TF with forcast (2017–2018)

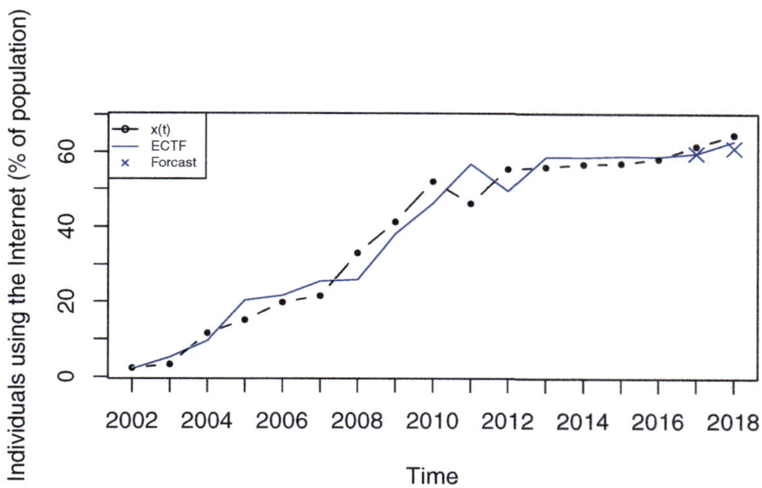

Fig. 3 Real data and estimated CTF with forcast (2017–2018)

6.1 Comparing the Fitness of the Proposed Model and the Lognormal Model

In this section, we compare the performance of the Generalized Goel-Okumoto curve-based Diffusion Process (GGODP) and the stochastic Lognormal Diffusion Process (LDP). Since GGODP is an extension of LDP, we assess the accuracy of these two models by examining the Relative Absolute Error (RAE), Mean Absolute Error

Table 3 Goodness of fit of the two models

Measures	Values of stochastic GGODP	Values of LDP
RAE	0.1147888	1.409076
MAE	2.207508	27.09798
SMAPE	0.08060349	1.02502

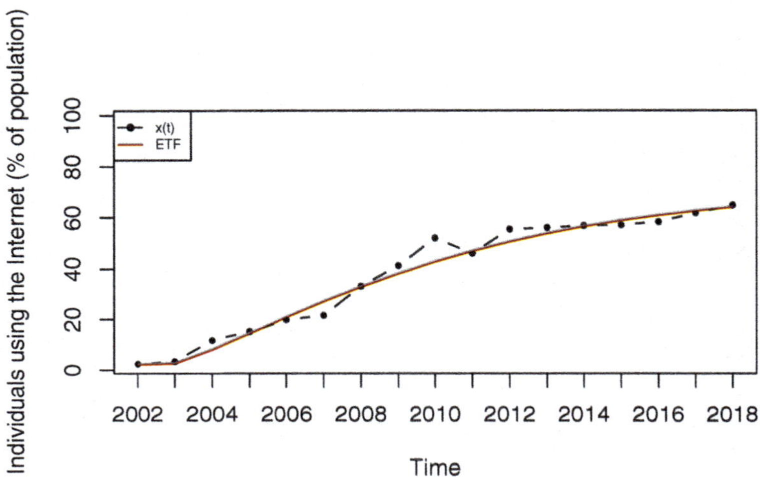

Fig. 4 The real data (Individuals using the Internet in Morocco) versus those fitted by the stochastic GGODP

(MAE), and Symmetric Mean Absolute Percentage Error (SMAPE) metrics. The results, presented in Table 3, allow us to evaluate the outcomes obtained from the proposed diffusion process.

A comparison was made between the results obtained using the stochastic GGODP and the stochastic LDP. This comparison is illustrated in Figs. 4 and 5, which depict the outcomes obtained from each model.

These figures show that the stochastic GGODP was more suitable than the stochastic LDP.

7 Conclusion

- In this research on the novel stochastic diffusion process utilizing the Generalized Goel-Okumoto curve, our initial focus is on determining the probabilistic characteristics of the process. To estimate its parameters, we perform a discrete sampling of the process using the MLM. The resulting equation from maximizing

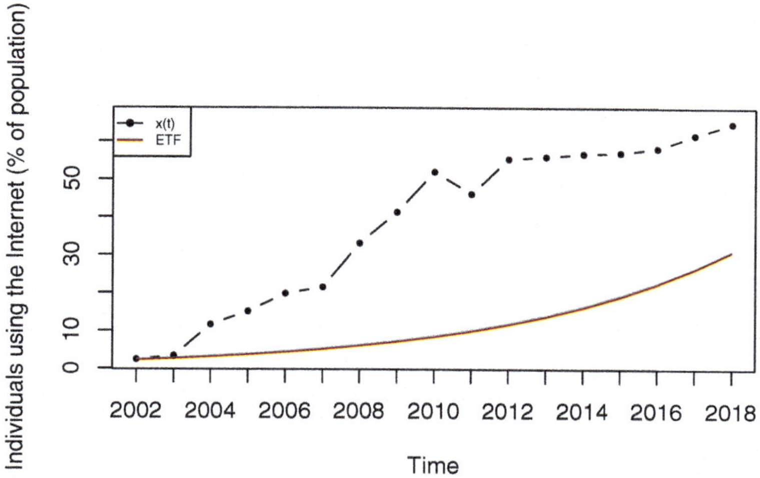

Fig. 5 The real data (Individuals using the Internet in Morocco) versus those fitted by the stochastic Lognormal Diffusion Process (LDP)

the log-likelihood function is subsequently solved using the Simulated Annealing method.

- The stochastic diffusion process based on the Generalized Goel-Okumoto curve was employed to analyze the percentage of individuals using the Internet in Morocco. This approach provided an improved representation of the observed time series data from 2002 to 2017 and yielded enhanced short-term forecasts for the period of 2017 to 2018. The results obtained, as shown in Table 1 and Figs. 2 and 3, lead us to the conclusion that when the proposed model is applied to the real-world case using the estimation methodology outlined in Sect. 4, the fitting and predictions achieved based on both the ETF and the ECTF exhibit a high level of accuracy, as indicated in Table 2.
- An intriguing avenue for future investigation would involve exploring the potential for constructing a nonhomogeneous generalized Goel-Okumoto model by incorporating exogenous factors into the drift component, similar to the approach employed in other diffusion models (e.g., [10, 14]). This extension would allow for the analysis of factors influencing the dynamics of Individuals using the Internet (% of population), such as education, GDP per capita, and the consumer price index.
- We evaluated three error measurements, RAE, MAE and SMAPE, in order to compare the forecasting precision of the two models. For these error measures, the values obtained revealed that the stochastic GGODP was more accurate than the stochastic LDP.

Acknowledgements The authors are very grateful to the Editor and referees.

References

1. Artzrouni, M., Reneke, J.: Stochastic differential equations in mathematical demography: a review. Appl. Math. Comput. **39**(3), 139–153 (1990)
2. Brousseau, E., Curien, N.: Internet economics, digital economics. Internet and Digital Economics Principles, Methods and Applications, vol. 1 (2007)
3. Capocelli, R.M., Ricciardi, L.M.: A diffusion model for population growth in random environment. Theor. Popul. Biol. **5**(1), 28–41 (1974)
4. David, P.A.: The dynamo and the computer: an historical perspective on the modern productivity paradox. Am. Econ. Rev. **80**(2), 355–361 (1990)
5. David, P.A.: Understanding digital technology's evolution and the path of measured productivity growth: present and future in the mirror of the past. In: Understanding the Digital Economy: Data, Tools, and Research. MIT, Cambridge, MA (2000)
6. Giovanis, A.N., Skiadas, C.H.: A stochastic logistic innovation diffusion model studying the electricity consumption in Greece and the United States. Technol. Forecast. Soc. Chang. **61**(3), 235–246 (1999)
7. Goel, A.L.: Software reliability models: assumptions, limitations, and applicability. IEEE Trans. Softw. Eng. **12**, 1411–1423 (1985)
8. Goel, A.L., Okumoto, K.: Time-dependent error-detection rate model for software reliability and other performance measures. IEEE Trans. Reliab. **28**(3), 206–211 (1979)
9. Goodwin, P., Lawton, R.: On the asymmetry of the symmetric MAPE. Int. J. Forecast. **15**(4), 405–408 (1999)
10. Gutiérrez, R., Gutiérrez-Sánchez, R., Nafidi, A.: Electricity consumption in Morocco: stochastic Gompertz diffusion analysis with exogenous factors. Appl. Energy **83**(10), 1139–1151 (2006)
11. Kirkpatrick, S., Gelatt, C.D., Vecchi, M.P.: Optimization by simulated annealing. Science **220**(4598), 671–680 (1983)
12. Kloeden, P.E., Platen, E.: Numerical Solution of Stochastic Differential Equations, vol. 23. Springer Science & Business Media (2013)
13. Nafidi, A., Bahij, M., Gutiérrez-Sánchez, R., Achchab, B.: Two-parameter stochastic weibull diffusion model: statistical inference and application to real modeling example. Mathematics **8**(2), 160 (2020)
14. Nafidi, A., Gutiérrez, R., Gutiérrez-Sánchez, R., Ramos-Ábalos, E., El Hachimi, S.: Modelling and predicting electricity consumption in Spain using the stochastic Gamma diffusion process with exogenous factors. Energy **113**, 309–318 (2016)
15. Nafidi, A., Rida, O., Achchab, B.: Stochastic diffusion process based on generalized brody curve: application to real data. J. Math. Stat. Stud. **2**(1), 01–11 (2021)
16. Román-Román, P., Torres-Ruiz, F.: A stochastic model related to the Richards-type growth curve. Estimation by means of simulated annealing and variable neighborhood search. Appl. Math. Comput. **266**, 579–598 (2015)
17. Seybert, H.: Internet use in households and by individuals in 2011. Eurostat Stat. Focus **66**, 2011 (2011)
18. Tuckwell, H.C., Koziol, J.A.: Logistic population growth under random dispersal. Bull. Math. Biol. **49**(4), 495–506 (1987)
19. Vera, J.F., Dìaz-Garcìa, J.A.: A global simulated annealing heuristic for the three-parameter log-normal maximum likelihood estimation. Comput. Stat. Data Anal. **52**(12), 5055–5065 (2008)
20. Zehna, P.W.: Invariance of maximum likelihood estimators. Ann. Math. Stat. **37**(3), 744 (1966)

A Comparative Study of Dam-Break Problem over a Sandy Bottom by an Unstructured Finite Volume Method

Sanae Jelti

Abstract The study described in this work focuses on the dam-break problem over a sandy bed. The goal is to analyze the effects and reactions of various parameters involved in the problem. The problem is mathematically modeled using a coupled model and a non-capacity model. To solve the mathematical model numerically, an unstructured finite volume method is employed. This method allows for the discretization of the problem domain into a mesh of cells, where the conservation equations are solved at the cell level. In order to achieve second-order accuracy in both space and time, the MUSCL method is used for spatial discretization, while the Runge–Kutta method is used for time integration. One particular aspect of the numerical implementation is the treatment of the source term, which is handled using an original approach. This treatment ensures the accurate representation of the physical phenomena involved in the dam-break problem. To enhance the accuracy of the results and reduce computational time, an adaptive mesh is employed. This means that the mesh is dynamically refined or coarsened in regions of interest based on certain criteria, allowing for a higher level of accuracy in those areas while saving computational resources in less critical regions. The study considers several cases, likely involving different initial conditions, boundary conditions, or parameter values. The results obtained from these simulations are presented and analyzed, highlighting the differences observed for different computational times.

1 Introduction

The treatment of complex problems, such as dam break coupled with sediment transport and the resulting extensive damages, requires careful consideration. Numerical modeling is commonly employed to investigate these types of physical phenomena. Based on previous studies [4, 6, 8, 9, 11, 14], it has been observed that in unsteady flows involving highly concentrated sediment debris, the evolution of the bed is

S. Jelti (✉)
Mechanic and Energetic Laboratory, FSO, University Mohammed First, 60050 Oujda, Morocco
e-mail: s.jelti@ump.ac.ma

S. Melliani et al. (eds.), *Applied Mathematics and Modelling in Finance, Marketing and Economics*, Studies in Computational Intelligence 1114,
https://doi.org/10.1007/978-3-031-42847-0_15

significantly more pronounced compared to the evolution of the water free-surface. Consequently, our current study, described in [4, 5], utilizes a coupled model consisting of shallow water equations for a water-sediment mixture, along with a non-capacity model.

The mathematical formulation of the problem consists of five equations: the shallow-water equations for a water-sediment mixture in a two-dimensional scenario, coupled with the transport diffusion equation for sediment particles and the equation for bed morphology change. To solve the mathematical model numerically, an unstructured finite volume method based on the Roe scheme, introduced by Roe [10], is employed. The source term is discretized using a novel method developed by Jelti et al. [9], which satisfies the C-property. To achieve second-order accuracy in both space and time, the MUSCL method with a generalized minmod limiter is utilized, along with the Runge–Kutta method. Furthermore, a local mesh refinement technique is implemented, using sediment concentration as a monitoring function, to attain a higher level of accuracy while optimizing computational costs. Special attention is given to examining the influence of water height on the flow behavior.

Our focus is on studying the evolution of flow behavior and bed changes in a two-dimensional dam-break problem. Through the numerical scheme employed, we have demonstrated the capability of accurately and efficiently simulating the shallow-water equations for a water-sediment mixture, coupled with a non-capacity model that incorporates the mass conservation equation for total sediment load and the equation governing bed morphological changes. The results obtained highlight the effectiveness of the numerical scheme in capturing the dynamics of the system.

The remaining sections of this paper are organized as follows. In Sect. 2, we introduce the governing equations that describe the problem. Section 3 presents the unstructured finite volume Roe scheme, discussing its second-order accuracy in space and time. We also describe the boundary conditions employed in the simulations and present the novel discretization approach for the source term, ensuring the satisfaction of the C-property. Additionally, the procedure for mesh refinement is explained. In Sect. 4, we present the numerical tests conducted and the corresponding results. Finally, in Sect. 5, we provide concluding remarks summarizing the main findings of the study.

2 Mathematical Model

In this study, we investigate a two-dimensional flow in a channel with a constant-width rectangular cross-section, where the channel bed is comprised of sandy material. The mathematical formulation we present here is built upon the shallow-water equations for a mixture of sediment and water, coupled with a non-capacity model that includes equations for sediment mass conservation and bed rate change. Notably, the same mathematical model has been employed in previous works [8, 9, 11, 14]. The governing equations for the system can be expressed as follows:

$$\frac{\partial h}{\partial t}+\frac{\partial (hu)}{\partial x}+\frac{\partial (hv)}{\partial y}=\frac{E-D}{1-p} \tag{1}$$

$$\frac{\partial (hu)}{\partial t}+\frac{\partial (hu^2+\frac{1}{2}gh^2)}{\partial x}+\frac{\partial (huv)}{\partial y}=B_x \tag{2}$$

$$\frac{\partial (hv)}{\partial t}+\frac{\partial (huv)}{\partial x}+\frac{\partial (hv^2+\frac{1}{2}gh^2)}{\partial y}=B_y \tag{3}$$

$$\frac{\partial (hc)}{\partial t}+\frac{\partial (huc)}{\partial x}+\frac{\partial (hvc)}{\partial y}=E-D \tag{4}$$

$$\frac{\partial z}{\partial t}=-\frac{E-D}{1-p} \tag{5}$$

where B_x and B_y are the source terms defined by:

$$B_x=-gh\frac{\partial z}{\partial x}-\frac{\rho_s-\rho_w}{2\rho}gh^2\frac{\partial c}{\partial x}-ghS_{f_x}-\frac{\rho_0-\rho}{\rho}\frac{E-D}{1-p}u \tag{6}$$

$$B_y=-gh\frac{\partial z}{\partial y}-\frac{\rho_s-\rho_w}{2\rho}gh^2\frac{\partial c}{\partial y}-ghS_{f_y}-\frac{\rho_0-\rho}{\rho}\frac{E-D}{1-p}v \tag{7}$$

In the given system, several variables are defined as follows: t represents time, x and y indicate horizontal coordinates, h denotes the flow depth, u and v are the velocities averaged over the depth in the x- and y-directions respectively, z represents the elevation of the bed. Additionally, c represents the concentration of sediment averaged over the flux, g symbolizes the acceleration due to gravity, and p represents the porosity of the bed sediment. The variables D and E represent the fluxes of sediment deposition and entrainment, respectively. These fluxes occur at the bottom boundary, signifying the exchange between the water column and the bed. The symbols S_{fx} and S_{fy} represent the friction slopes in the x- and y-directions respectively. The density of the water-sediment mixture is given by $\rho=\rho_w(1-c)+\rho_s c$, where ρ_w and ρ_s are the densities of water and sediment respectively. Furthermore, ρ_0 is defined as the density of the saturated bed, calculated as $\rho_w p+\rho_s(1-p)$.

The empirical functions considered in this paper are the same as [9]. The friction slopes S_{fx} and S_{fy} are defined using the Manning roughness coefficient n_b, as

$$S_{fx}=\frac{n_b^2}{h^{\frac{4}{3}}}u\sqrt{u^2+v^2} \tag{8}$$

$$S_{fy}=\frac{n_b^2}{h^{\frac{4}{3}}}v\sqrt{u^2+v^2} \tag{9}$$

For deposition D, the relation used is:

$$D = \omega(1 - c_a)^m c_a \tag{10}$$

In the given context, the exponent m is assigned a value of 2. Additionally, c_a represents the local near-bed sediment concentration in terms of volume. It is assumed to be directly proportional to the depth-averaged concentration, denoted by c. Mathematically, this relationship can be expressed as $c_a = \alpha c$, where α is an empirical coefficient that is typically greater than one [13].

$$\alpha = \min\left(2, \frac{1-p}{c}\right) \tag{11}$$

ω is the settling velocity of sediment particle in tranquil water [12, 13]:

$$\omega = 1.1\sqrt{\left(\frac{\rho_s}{\rho} - 1\right) g\, d} \tag{12}$$

where d is the diameter of the sediment grain. In our case d is larger than 1 mm.

For the entrainment, we use:

$$E = \begin{cases} \varphi \dfrac{\theta - \theta_c}{h} \dfrac{\sqrt{u^2 + v^2}}{d^{0.2}} & if \quad \theta \geq \theta_c \\ 0 & otherwise \end{cases} \tag{13}$$

where the variable φ represents a coefficient used to regulate the erosion force. It is assigned a value of 0.015 $m^{1.2}$. The critical value of Shield's parameter required for the initiation of sediment motion is denoted by θ_c, which is set to 0.045. Additionally, θ represents the Shield's coefficient, defined by [4, 11].

$$\theta = \frac{u_*^2}{gd\sqrt{\frac{\rho_s}{\rho_w} - 1}} \tag{14}$$

u_* is the friction velocity defined by:

$$u_*^2 = \sqrt{\frac{f}{8}} \cdot \left|\sqrt{u^2 + v^2}\right| \tag{15}$$

where f is the Darcy–Weisbach friction factor defined by [11]:

$$f = \frac{8\, g\, n_f^2}{h^{1/3}}. \tag{16}$$

3 Numerical Scheme

In this section of the study, we introduce the approximation scheme utilized along with the methods employed to attain higher order accuracy. Equations (1)–(5) can be expressed in a conservative form as shown below:

$$\frac{\partial U}{\partial t} + \frac{\partial F(U)}{\partial x} + \frac{\partial G(U)}{\partial y} = S(U) + Q(U) \tag{17}$$

where

$$U = \begin{pmatrix} h \\ hu \\ hv \\ hc \\ z \end{pmatrix}, \quad F = \begin{pmatrix} hu \\ hu^2 + \frac{1}{2}gh^2 \\ huv \\ huc \\ 0 \end{pmatrix}, \quad G = \begin{pmatrix} hv \\ huv \\ hv^2 + \frac{1}{2}gh^2 \\ hvc \\ 0 \end{pmatrix}, \tag{18}$$

$$S = \begin{pmatrix} 0 \\ -gh\frac{\partial z}{\partial x} - \frac{(\rho_s-\rho_w)}{2\rho}gh^2\frac{\partial c}{\partial x} \\ -gh\frac{\partial z}{\partial y} - \frac{(\rho_s-\rho_w)}{2\rho}gh^2\frac{\partial c}{\partial y} \\ 0 \\ 0 \end{pmatrix} \text{and} Q = \begin{pmatrix} \frac{E-D}{1-p} \\ -ghS_{f_x} - \frac{\rho_0-\rho}{\rho}\frac{E-D}{1-p}u \\ -ghS_{f_y} - \frac{\rho_0-\rho}{\rho}\frac{E-D}{1-p}v \\ E-D \\ -\frac{E-D}{1-p} \end{pmatrix}. \tag{19}$$

3.1 Finite Volume Roe-Scheme

The mathematical model described by Eq. (17) will be discretized on an unstructured grid using the finite volume Roe scheme [10]. In this study, only triangular grids are considered. Consequently, we divide the time interval into sub-intervals $[t_n, t_{n+1}]$ with a step size of Δt, and discretize the spatial domain into conforming triangular elements T_i. Each triangle represents a control volume, and the variables are located at the geometric centers of the cells. Therefore, applying a finite-volume discretization to Eq. (17) yields:

$$U_i^{n+1} = U_i^n - \frac{\Delta t}{|T_i|}\sum_{j\in N(i)}\int_{\Gamma_{ij}} \mathscr{F}(U^n, \overrightarrow{\eta_{ij}})d\Gamma + \frac{\Delta t}{|T_i|}\int_{Ti} S(U^n)d\Omega + \frac{\Delta t}{|T_i|}\int_{Ti} Q(U^n)d\Omega \tag{20}$$

where:

- $$\mathscr{F}(\mathscr{U}, \overrightarrow{\eta_{ij}}) = \mathscr{F}(\mathscr{U})\eta + \mathscr{G}(\mathscr{U})\eta \tag{21}$$

- $N(i)$ is the set of neighboring triangles of the cell T_i, U_i^n is an averaged value of the solution U in the cell T_i at time t_n:

$$U_i^n = \frac{1}{|T_i|}\int_{T_i} U^n d\Omega \tag{22}$$

- $|T_i|$ denote the area of T_i
- Γ_{ij} is the interface between the two control volumes T_i and T_j
- $\overrightarrow{\eta_{ij}} = (n_x, n_y)^T$ denote the unit outward normal to Γ_{ij}.

The numerical flux is defined by Φ_{ij} such as:

$$\int_{\Gamma_{ij}} \mathscr{F}(U^n, \overrightarrow{\eta_{ij}})d\Gamma = \Phi_{ij}\left|\Gamma_{ij}\right| \tag{23}$$

Equation (20) becomes:

$$U_i^{n+1} = U_i^n - \frac{\Delta t}{|T_i|}\sum_{j\in N(i)} \Phi_{ij}|\Gamma_{ij}| + \frac{\Delta t}{|T_i|}\int_{T_i} S(U^n)d\,\Omega + \frac{\Delta t}{|T_i|}\int_{T_i} Q(U^n)d\,\Omega, \tag{24}$$

Using Roe scheme, the numerical flux is defined as

$$\Phi_{ij}(U_{ij}^L, U_{ij}^R) = \frac{1}{2}(\mathscr{F}(U_{ij}^L) + \mathscr{F}(U_{ij}^R)) - \frac{1}{2}\left|\mathscr{A}(\widetilde{U}_{ij}^n)\right|(U_{ij}^R - U_{ij}^L), \tag{25}$$

U_{ij}^R, U_{ij}^L are the right and left approximations of the solution U at the interface Γ_{ij} at time t_n.

The Jacobian matrix $\mathscr{A}(\widetilde{U}_{ij}^n)$ uses $\widetilde{U}_{ij}^n$ which is the Roe average defined as

$$\widetilde{U}_{ij}^n = \begin{pmatrix} \frac{h_i+h_j}{2} \\ \frac{h_i+h_j}{2}\left(\frac{u_i\sqrt{h_i}+u_j\sqrt{h_j}}{\sqrt{h_i}+\sqrt{h_j}}\eta_x + \frac{v_i\sqrt{h_i}+v_j\sqrt{h_j}}{\sqrt{h_i}+\sqrt{h_j}}\eta_y\right) \\ \frac{h_i+h_j}{2}\left(-\frac{u_i\sqrt{h_i}+u_j\sqrt{h_j}}{\sqrt{h_i}+\sqrt{h_j}}\eta_y + \frac{v_i\sqrt{h_i}+v_j\sqrt{h_j}}{\sqrt{h_i}+\sqrt{h_j}}\eta_x\right) \\ \frac{h_i+h_j}{2}\left(\frac{c_i\sqrt{h_i}+c_j\sqrt{h_j}}{\sqrt{h_i}+\sqrt{h_j}}\right) \\ \frac{z_i+z_j}{2} \end{pmatrix} \tag{26}$$

The Jacobian matrix $\mathscr{A}(\widetilde{U}_{ij}^n)$ is computed by considering the system defined in Eq. (17) without taking into account the term $Q(U)$.

$$\frac{\partial U}{\partial t} + \frac{\partial F(U)}{\partial x} + \frac{\partial G(U)}{\partial y} = S(U). \tag{27}$$

The system (27) is projected on the normal and on the tangential of the interface noted respectively by $\eta = (n_x, n_y)$ and $\tau = \eta^{\perp}$. Knowing that the normal and the tangential velocities are defined respectively as follows

$$U_\eta = un_x + vn_y \text{ and } U_\tau = -un_y + vn_x, \tag{28}$$

From the detailed calculation brought in [9], the Jacobian matrix $\mathscr{A}(\widetilde{U}_{ij}^n)$ is defined as

$$\mathscr{A}_\eta(U) = \begin{bmatrix} 0 & 1 & 0 & 0 & 0 \\ gh - U_\eta^2 - \frac{\rho_s - \rho_w}{2\rho} ghc & 2U_\eta & 0 & \frac{\rho_s - \rho_w}{2\rho} gh & gh \\ -U_\eta U_\tau & U_\tau & U_\eta & 0 & 0 \\ -U_\eta c & c & 0 & U_\eta & 0 \\ 0 & 0 & 0 & 0 & 0 \end{bmatrix}. \tag{29}$$

The discrete Eq. (24) is implemented using the decomposition procedure given in [3]:

$$U_i^* = U_i^n - \frac{\Delta t}{|T_i|} \sum_{j \in N(i)} \Phi_{ij} \, |\Gamma_{ij}| + \frac{\Delta t}{|T_i|} \int_{T_i} S(U^n) d\,\Omega, \tag{30}$$

$$U_i^{n+1} = U_i^* + \frac{\Delta t}{|T_i|} \int_{T_i} Q(U_i^*) d\,\Omega. \tag{31}$$

3.2 Second Order Approximation in Space

To achieve second-order accuracy in our finite volume method, we employ the Monotone Upstream-Centered Scheme for Conservation Laws (MUSCL) method, which incorporates a slope limiter in the spatial approximation. The MUSCL method discretization involves approximating the solution state U using linear interpolation at each cell interface Γ_{ij}. The left and right solution values at the interface Γ_{ij} are denoted as U_{ij}^L and U_{ij}^R respectively, and they are defined as follows:

$$U_{ij}^L = U_i + \frac{1}{2} \nabla U_i . \mathbf{X_i X_j} \text{ and } U_{ij}^R = U_j - \frac{1}{2} \nabla U_j . \mathbf{X_i X_j} \tag{32}$$

Where $X_i = (x_i, y_i)^T$ and $X_j = (x_j, y_j)^T$, are respectively, the barycenters coordinates of cells T_i and T_j. With $\mathbf{X_i X_j} = (x_j - x_i, y_j - y_i)$.

To ensure Total Variation Diminishing (TVD) property in our scheme, we introduce a slope limiter. Specifically, we apply the generalized Minmod limiter, which controls the gradients at each cell. The limited gradients are determined as follows:

$$\frac{\partial^{lim} U_i}{\partial x} = \frac{1}{2} \left[\min_{j \in V(i)} syn\left(\frac{\partial U_j}{\partial x} \right) + \max_{j \in V(i)} syn\left(\frac{\partial U_j}{\partial x} \right) \right] \min_{j \in V(i)} \left| \frac{\partial U_j}{\partial x} \right| \tag{33}$$

and

$$\frac{\partial^{lim} U_i}{\partial y} = \frac{1}{2}\left[\min_{j \epsilon V(i)} syn\left(\frac{\partial U_j}{\partial y}\right) + \max_{j \epsilon V(i)} syn\left(\frac{\partial U_j}{\partial y}\right)\right] \min_{j \epsilon V(i)} \left|\frac{\partial U_j}{\partial y}\right|. \quad (34)$$

3.3 Second Order Approximation in Time

To achieve second-order approximation in the temporal dimension, we employ the Runge–Kutta second-order scheme [7].

$$\begin{cases} U^* = U^n + \Delta t \ \mathscr{L}(U^n) \\ U^{**} = U^* + \Delta t \ \mathscr{L}(U^*) \\ U^{n+1} = \frac{1}{2}(U^n + U^{**}) \end{cases} . \quad (35)$$

3.4 Boundary Conditions

In this study, we adopt the same boundary conditions as presented in our previous work [9]. For open inflow and outflow boundaries, the flow variables are set to the same values as those at the interior of the flow. As for solid walls, the flow variables are mirrored at the corresponding boundary points, ensuring that the normal velocity component is zero at the boundary.

3.5 Treatment of the Source Term

The treatment of the source term in the governing Eqs. (1)–(5) follows the same approach developed in our previous work [9] for the two-dimensional case. It is important to note that the proposed discretization of the source term satisfies the C-property and incorporates data from both the left and right sides of the interface Γ_{ij}. Here, we directly provide a decomposition of the source term $\int_{T_i} S(U^n)$ presented in Eq. (24) as reported in [9]:

$$\int_{T_i} S(U^n) d\,\Omega = \frac{1}{2}(\mathscr{S}_i^R + \mathscr{S}_i^L) \quad (36)$$

where $\mathscr{S}_i^R$ and $\mathscr{S}_i^L$ are the right and left approximations of $\int_{T_i} S(U^n) d\,\Omega$

$$\mathscr{S}_i^R = \begin{pmatrix} 0 \\ I_{xi}^R - \frac{\rho_0-\rho_w}{2\rho} g\, h_i^2 \sum_{j\epsilon N(i)} c_{ij}^R\, n_{xij}|\Gamma_{ij}| \\ I_{yi}^R - \frac{\rho_0-\rho_w}{2\rho} g\, h_i^2 \sum_{j\epsilon N(i)} c_{ij}^R\, n_{yij}|\Gamma_{ij}| \\ 0 \\ 0 \end{pmatrix} \quad (37)$$

and

$$\mathscr{S}_i^L = \begin{pmatrix} 0 \\ I_{xi}^L - \frac{\rho_0-\rho_w}{2\rho} g\, h_i^2 \sum_{j\epsilon N(i)} c_{ij}^L\, n_{xij}|\Gamma_{ij}| \\ I_{yi}^L - \frac{\rho_0-\rho_w}{2\rho} g\, h_i^2 \sum_{j\epsilon N(i)} c_{ij}^L\, n_{yij}|\Gamma_{ij}| \\ 0 \\ 0 \end{pmatrix}. \quad (38)$$

A centred disretization is proposed for the term $\int_{T_i} Q(U^n)d\,\Omega$ is presented as:

$$\frac{1}{|T_i|}\int_{T_i} Q(U^n)d\,\Omega = \begin{pmatrix} \frac{E-D}{1-p} \\ -gh_i S_{f_x} - \frac{\rho_0-\rho}{\rho}\frac{E-D}{1-p}u_i \\ -gh_i S_{f_y} - \frac{\rho_0-\rho}{\rho}\frac{E-D}{1-p}v_i \\ E-D \\ -\frac{E-D}{1-p} \end{pmatrix}. \quad (39)$$

3.6 *Mesh Adaptation*

In this phase of the study, we employ mesh adaptation techniques to construct a nearly optimal mesh that can effectively capture small-scale hydraulic features. This approach allows us to avoid the need for excessively fine grids in smooth regions that are far from hydraulic jumps, as suggested in [1]. Similar to our previous work in [9], we utilize this method to enhance the efficiency of our scheme. To achieve this objective, we introduce an error indicator based on the gradient of the sediment concentration. This indicator only requires information from solution values within a single element at a time, making it computationally efficient. The indicator is evaluated using the following expression, as outlined in [2]:

$$\mathscr{C}_{T_i}^n = \frac{||\nabla(c(T_i))||}{max_{T_j}||\nabla(c(T_j))||}. \quad (40)$$

In the given expression, $\nabla(c(T_i))$ represents the Euclidean norm of the gradient of the sediment concentration c on the triangle T_i. The normalization of this quantity has the advantage that the criterion (40) is known to lie within the interval [0, 1].

4 Numerical Results

In this study, we focus on a square reservoir with a flat sandy bottom. The length and width of the reservoir are both set to 200 m. The dam has a thickness of 4 m, and the breach is assumed to be 75 m wide. The primary objective of this research is to investigate the influence of water height on the flow behavior and bed rate change. We conduct experiments using the partial dam break scenario with three different water heights. Additionally, we utilize sediment with three different diameters resembling sand, namely 1, 2, and 4 mm.

4.1 Test 1

The initial conditions are given by

$$\begin{aligned}
Z(x, y, 0) &= 0\,\mathrm{m} \\
u(x, y, 0) &= v(x, y, 0) = 0\,\mathrm{m/s} \\
h(x, y, 0) &= \begin{cases} 10\,\mathrm{m}, & \texttt{if} \quad x \leq 100\,\mathrm{m} \\ 1\,\mathrm{m}, & \texttt{otherwise} \end{cases} \\
c(x, y, 0) &= \begin{cases} 0.01, & \texttt{if} \quad x \leq 100\,\mathrm{m} \\ 0, & \texttt{otherwise} \end{cases}.
\end{aligned}$$

4.2 Test 2

The initial conditions are given by

$$\begin{aligned}
Z(x, y, 0) &= 0\,\mathrm{m} \\
u(x, y, 0) &= v(x, y, 0) = 0\,\mathrm{m/s} \\
h(x, y, 0) &= \begin{cases} 10\,\mathrm{m}, & \texttt{if} \quad x \leq 100\,\mathrm{m} \\ 3\,\mathrm{m}, & \texttt{otherwise} \end{cases} \\
c(x, y, 0) &= \begin{cases} 0.01, & \texttt{if} \quad x \leq 100\,\mathrm{m} \\ 0, & \texttt{otherwise} \end{cases}.
\end{aligned}$$

4.3 Test 3

The initial conditions are given by

$$
\begin{aligned}
Z(x, y, 0) &= 0\,\mathrm{m} \\
u(x, y, 0) &= v(x, y, 0) = 0\,\mathrm{m/s} \\
h(x, y, 0) &= \begin{cases} 10\,\mathrm{m}, & \texttt{if } x \leq 100\,\mathrm{m} \\ 5\,\mathrm{m}, & \texttt{otherwise} \end{cases} \\
c(x, y, 0) &= \begin{cases} 0.01, & \texttt{if } x \leq 100\,\mathrm{m} \\ 0, & \texttt{otherwise} \end{cases}
\end{aligned}
$$

Notice: All figures represent a cross section on $y = 125$.

Figure 1 illustrates a three-dimensional perspective of the dam-break scenario accompanied by the corresponding mesh adaptation at different calculation times. The simulation uses a sand material with a diameter of $d = 1$ mm. It is evident from the figure that the relief of the flow is very smooth, resulting in a clear and accurate solution. Moreover, the adaptive mesh follows the variations of the error indicator for the concentration gradient (40). After the dam-break event, the refinement is primarily located around the initial position of the dam and gradually extends over a larger area.

Figures 2 and 3 present a comparison of water free surface, bed, concentration, and velocity profiles at different times for various sediment diameters using a sand material with a diameter of $d = 1$ mm. As highlighted in previous studies [8, 9], the diameter of the sediment plays a significant role in the flow behavior and erosion dynamics. Based on the observations from Figs. 2 and 3, it can be concluded that the hydraulic jump is more pronounced when the sediment diameter is larger. Additionally, the erosion rate is more significant when the bed consists of finer sediment, resulting in higher concentration levels. Regarding the velocity profiles, higher values are observed for larger sediment sizes. It is important to note that the behavior described above is specific to the studied sand material, which is characterized as non-cohesive soil with weak physical cohesion between sand particles. Results may differ significantly when considering other types of soil with the same diameter, such as organic soil.

The influence of the dam's extent on flow behavior and the severity of damages is evidently significant. In this study, we specifically emphasize the impact of water height on dam-break flow. We conduct a comparative analysis of the bed rate change, concentration profiles, and velocity profiles, depicted in Figs. 4, 5, and 6, respectively, for various water height scenarios in the same dam-break event. Alongside the sediment diameter, the water height plays a remarkable role in shaping the dynamics of the dam-break flow.

- From the bed evolution profiles presented in Fig. 4, it is evident that as the water height increases, the depth of erosion becomes more significant. In other words,

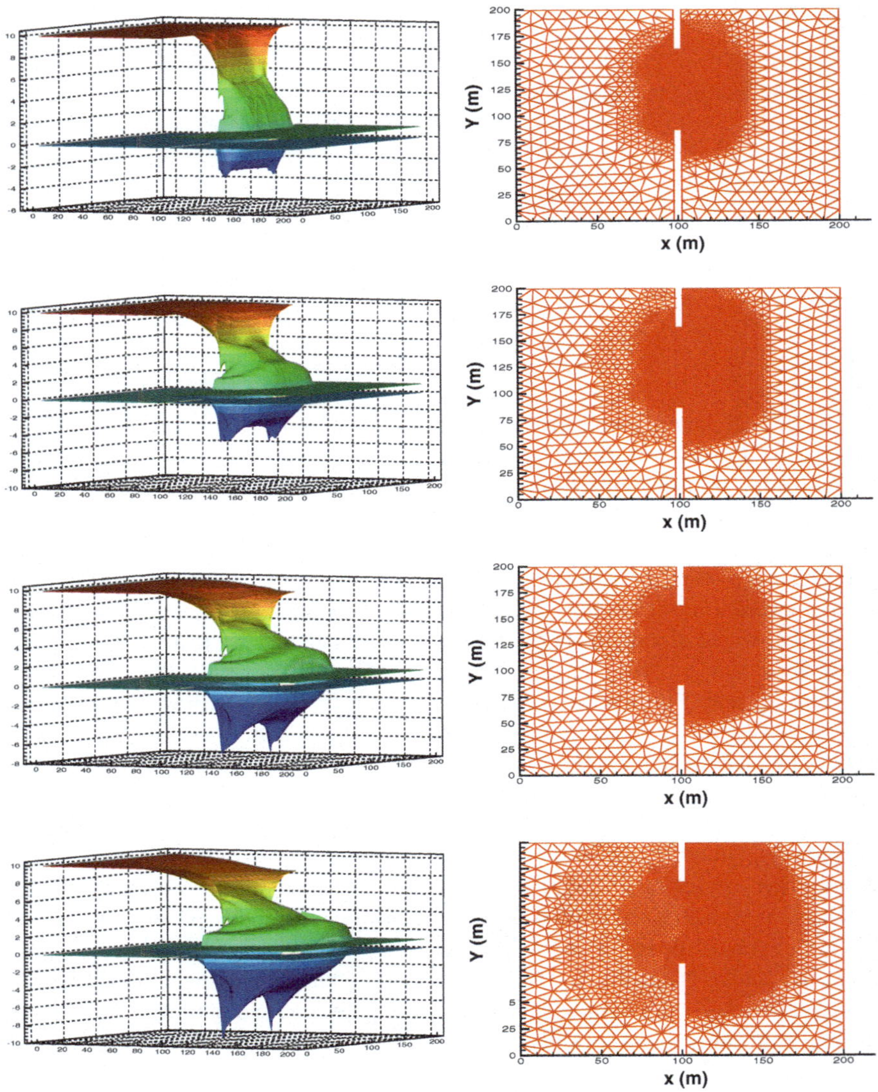

Fig. 1 Water free surface and bed profiles (first column), mesh adaptation (second column) at different time using $d = 1\,\mathrm{mm}$ for Test 1 (from top to bottom: $t = 2, 4, 6, 8\,\mathrm{s}$)

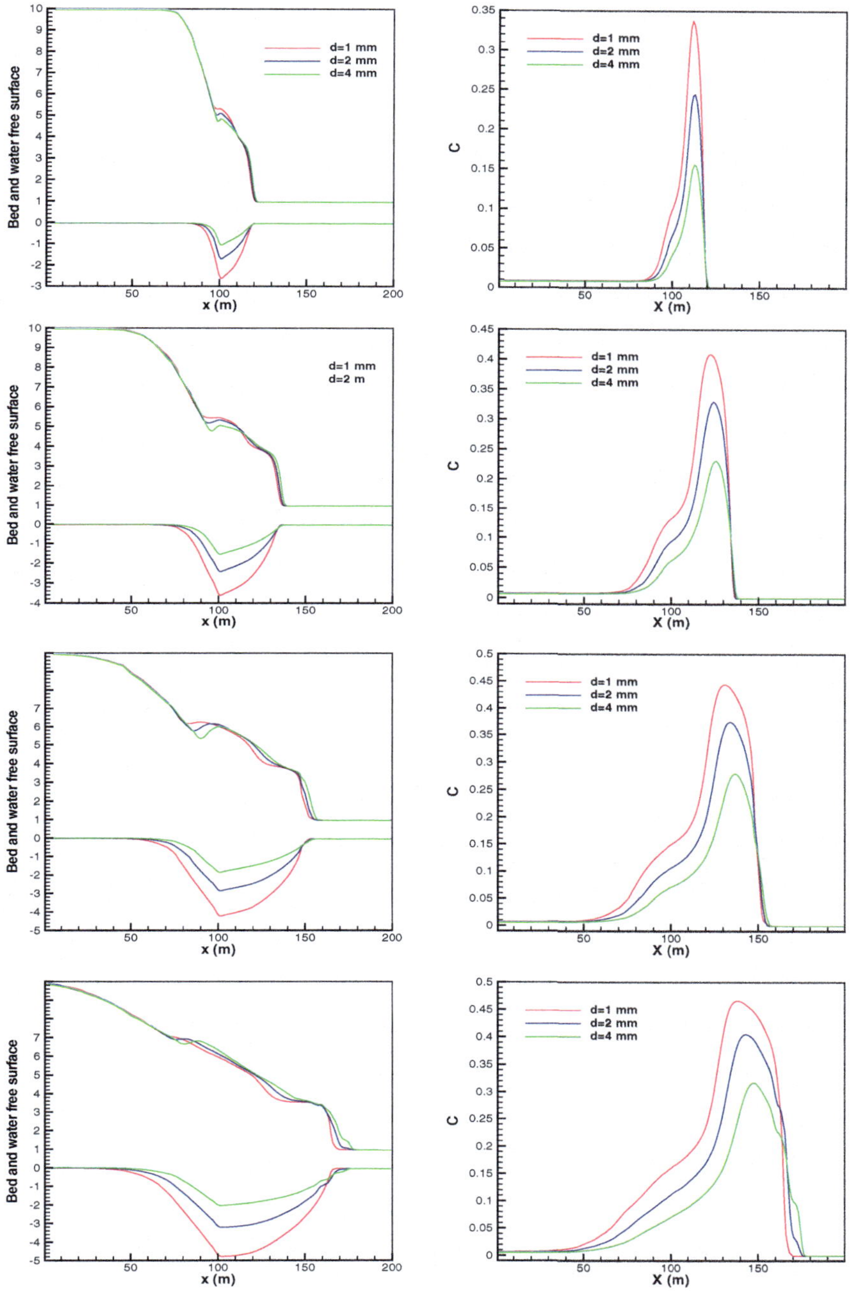

Fig. 2 Water free surface and bed profiles with their corresponding concentrations for different sizes of sediment at different time for Test 1 (from top to bottom: $t = 2, 4, 6, 8$ s)

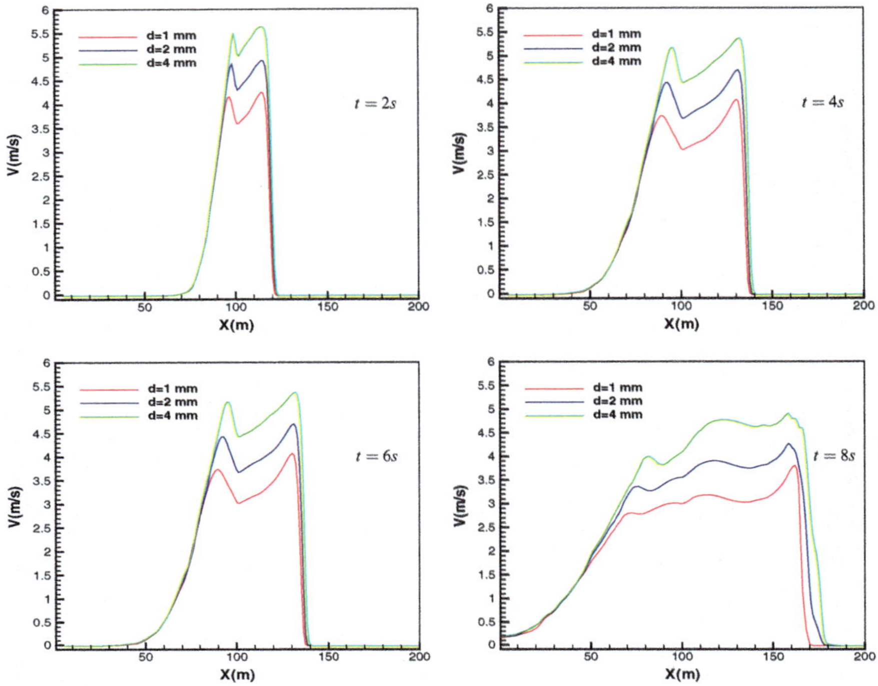

Fig. 3 Velocity profiles for different sizes of sediment at different time for Test 1 (t = 2, 4, 6, 8 s)

higher water levels result in more substantial erosion, leading to deeper changes in the bed morphology.

- From Fig. 5 we remark on velocity profiles that the higher concentrations backs to test 1 namely the of higher water level.
- By examining the velocity profiles depicted in Fig. 5, we observe that higher sediment concentrations are associated with higher water levels, particularly in the case of test 1. This suggests that as the water level increases, the sediment concentration in the flow also increases, resulting in a corresponding impact on the velocity profiles.

5 Conclusion

This paper presents a comparative study of dam-break flow over a sandy bottom, focusing on the impact of different parameters on this type of hydraulic phenomenon. The mathematical model employed comprises the two-dimensional shallow-water equations for the water-sediment mixture, the transport diffusion equation for sediment particles, and the equation for bed morphology changes. To numerically solve

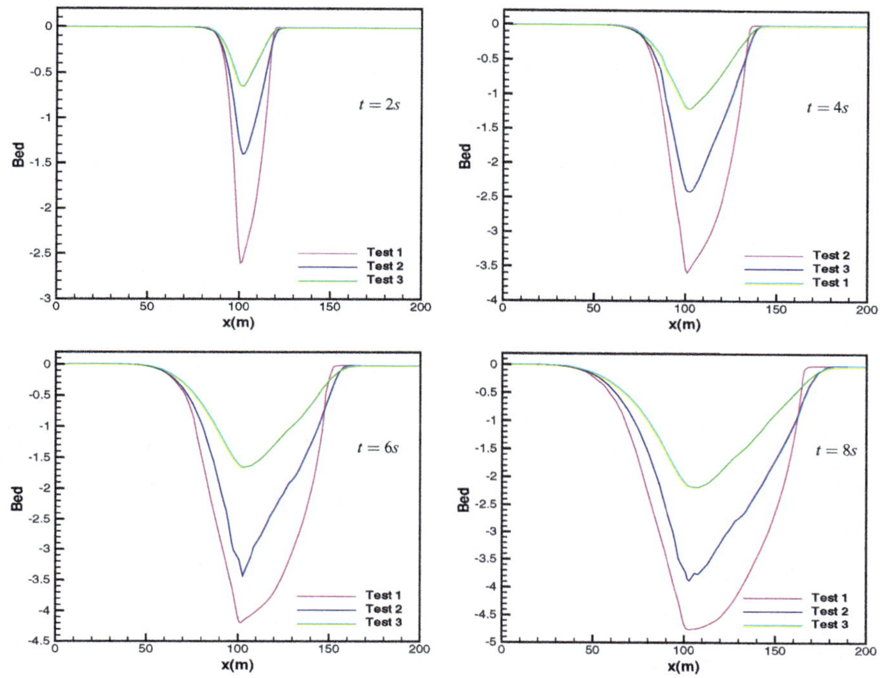

Fig. 4 A comparison of bed profiles at different time using $d = 1$ mm for different tests

the system, a finite volume method, specifically the Roe scheme, is utilized on unstructured grids. To achieve second-order accuracy in both space and time, a minmood limiter and the Runge–Kutta method are employed. The source term is discretized using a special technique that ensures the satisfaction of the C-property, as previously developed in Jelti et al. [9]. Additionally, mesh adaptation is implemented to enhance computational efficiency and accuracy, utilizing the gradient of sediment concentration as an error indicator.

In this study, we conducted a comparison of results from different cases of the same dam-break problem. Based on the findings, we can conclude that the employed scheme demonstrates a high level of accuracy and stability in addressing the considered physical problem. The obtained results validate the reliability and robustness of the scheme in accurately capturing the dynamics of the dam-break flow. We assume from the obtained results that:

- The utilization of shallow water equations for modeling water-sediment mixtures is crucial when dealing with scenarios involving a high concentration of sediment flux. This is primarily attributed to the significant rate of bed change compared to the evolution of the water-free surface. Additionally, the substantial amount of sediment transported by the flow necessitates the inclusion of sediment dynamics in the governing equations. By incorporating the shallow water equations, we can

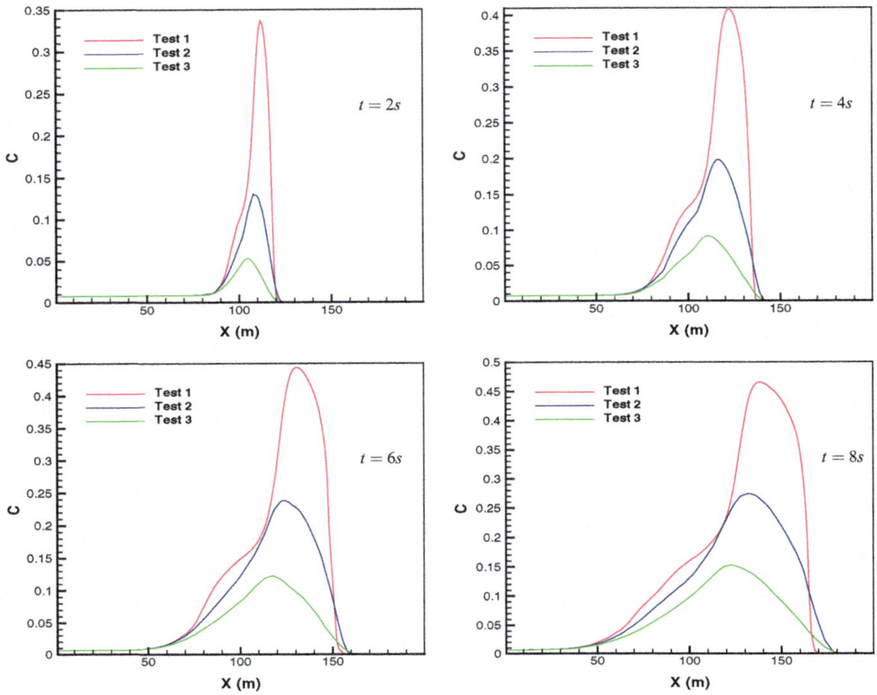

Fig. 5 A comparison of concentration profiles at different time using $d = 1$ mm for different tests

accurately capture the complex interactions between water and sediment, ensuring a comprehensive representation of the physical processes involved.

- In the context of unsteady flows such as the dam-break problem, it is essential to employ a non-capacity model. This is because the quantity of transported sediments varies both temporally and spatially. Using a capacity model, which assumes a fixed sediment capacity, may lead to an underestimation of the rate of bed change. By utilizing a non-capacity model, we can accurately capture the dynamic nature of sediment transport, accounting for the varying sediment concentrations and accurately predicting the rate of bed change over time and space. This approach ensures a more realistic representation of the dam-break flow and its associated sediment dynamics.
- The size of sediment particles has a significant impact on the erosion rate. Finer sediments tend to result in more substantial erosion, leading to higher sediment concentration profiles. On the water-free surface profiles, larger sediment particles produce more pronounced hydraulic jumps. This implies that the behavior of the dam-break flow, including erosion rates and hydraulic jumps, is influenced by the size of the sediment present. Therefore, careful consideration of sediment size is crucial for accurately predicting the dynamics and characteristics of the flow.

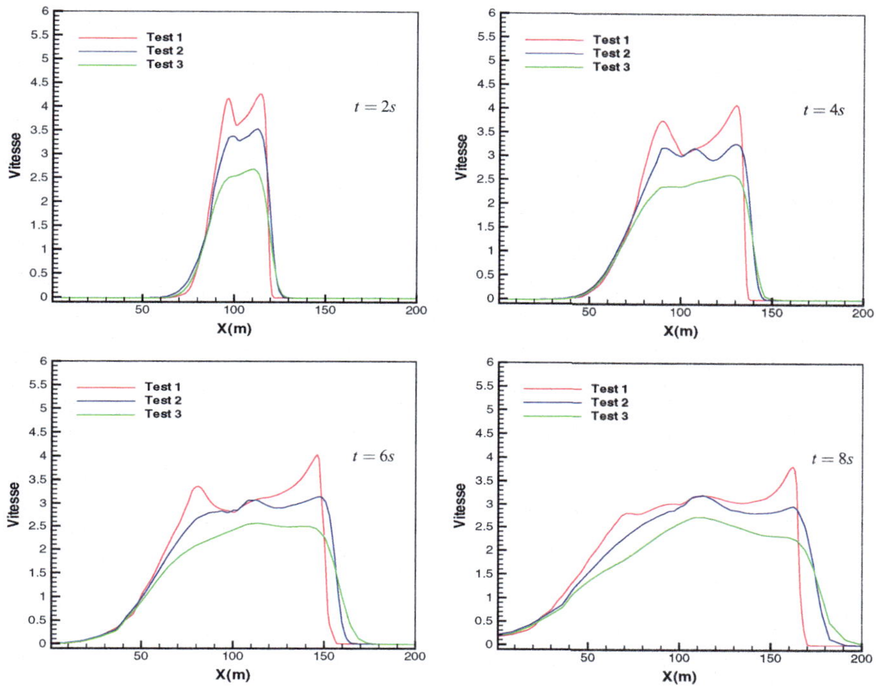

Fig. 6 A comparison of velocity profiles at different time using $d = 1$ mm for different tests

- The water height in a dam-break problem plays a significant role in determining the behavior of the flow. Higher water levels result in the erosion reaching deeper levels, which, in turn, directly impacts the concentration and velocity profiles. With increased water height, the erosion process becomes more pronounced, leading to higher sediment concentrations and altered velocity profiles. Therefore, the water height serves as a crucial parameter that influences the overall dynamics and characteristics of the dam-break flow.

This study focused on investigating the dynamics of dam-break flow over a non-cohesive sandy bottom. However, future research endeavors will be dedicated to exploring the behavior of dam-break flow over different types of soil, with a particular emphasis on cohesive sediment bottoms.

References

1. Benkhaldoun, F., Elmahi, I., Seaid, M.: A new finite volume method for flux-gradient and source-term blancing in shallow water equations. Comput. Methods Appl. Mech. Eng. **199**, 49–52 (2010)
2. Benkhaldoun, F., Elmahi, I., Sari, S., Seaid, M.: An unstructured finite-volume method for coupled models of suspended sediment and bed load transport in shallow-waterflows. Int. J. Numer. Methods Fluids **72**, 967–993 (2013)
3. Billet, S.J., Toro, E.F.: On WAF-type schemes for multidimentional hyperbolic conservation laws. J. Comput. Phys. **130**, 1–24 (1997)
4. Cao, Z., Pender, G., Wallis, S., Carling, P.: Computational dam-break hydraulics over erodible sediment bed. J. Hydraul. Eng. **130**, 389–703 (2004)
5. Cao, Z., Day, R., Egashira, S.: Coupled and decoupled numerical modelling of flow and morphological evolution in alluvial rivers. J. Hydraul. Eng. **128**, 306–321 (2002)
6. Chaojun, O., Siming, H., Qiang, X.: MacCormack-TVD finite difference solution for dam break hydraulics over erodible sediment beds. J. Hydraul. Eng. **372** (2014). https://doi.org/10.1061/(ASCE)HY.1943-7900.0000986
7. Gottlieb, S., Chi-Wang, S.: Total variation diminishing Runge-Kutta schemes. Math. Comput. **67**, 73–85 (1998)
8. Jelti, S., Mezouari, M., Boulerhcha, M.: Numerical modeling of dam-break flow over erodible bed by Roe scheme with an original discretization of source term. Int. J. Fluid Mech. Res. **45**, 21–36 (2017)
9. Jelti, S., Boulerhcha, M.: Numerical modeling of two dimensional non-capacity model for sediment transport by an unstructured finite volume method with a new discretization of source term. Math. Comput. Simul. **197**, 253–276 (2022)
10. Roe, P.L.: Approximate Riemann solvers, parameter vectors and difference schemes. J. Comput. Phys. **43**, 357–372 (1981)
11. Simpson, G., Castelltort, S.: Coupled model of surface water flow, sediment transport and morphological evolution. Comput. Geosci. **32**, 1600–1614 (2006)
12. Van Rijn, L.C.: Sediment transport, part II: suspended load transport. J. Sci. Comput. **110**, 1613–1641 (1984)
13. Weiming, Wu.: Computational River Dynamics. Taylor and Francis (2008)
14. Yue, Z., Cao, Z., Li, X., Che, T.: Two-dimensional coupled mathematical modeling of fluvial processes with intense sediment transport and rapid bed evolution. Sci. China Ser. G: Phys. Mech. Astron. **51**(9), 1427–1438 (2008)

Valuing a European Option Under the Heston Model with Interest Rate

Siham Bayad, Khalid Hilal, and Abdelmajid El Hajaji

Abstract In this research study, we derive a closed-form pricing formula for European options with analytical solution under the Heston model with the interest rate; in order to follow two-factor model by using the short-term interest rate and the volatility of the short term rate as the two factors. Heston-Longsraff-Schwartz hybrid model is proposed. Therefore, the numerical results in this paper represented different situations of computing European call option prices than can be more close to reality.

1 Introduction

One of the most critical ideas in modern financial theory is the Black Scholes model. Fischer Black, Robert Merton, and Myron Scholes developed it in 1973 [2], and it is still widely used today. It is known to be one of the easiest methods of determining fair prices of options. The model assumes that volatility is constant, wich contradict to the phenomen of "volatility smile" [1], Therefore stochastic models of volatility have been adopted and are especially common for derivatives pricing and hedging. For example, Johnson [6] and scott [10] used the Monte Carlo simulation technique to simulate the stochastic processes, While Wiggins [7] adopted the finite methode to solve the PDEs governing option prices. On the another hand, for situations in which stochastic volatility follow geometric Brownian motion Hull and With [8] calculated option prices in series form. In 1993 Heston [5] derived a colsed-form pricing formula for European options when stochastic volatility defined by the CIR (Cox-Ingersoll-Ross) model. Many studies recently incorporated the stochastic interest into the model of stochastic volatility to form a hybrid model. Grzelak and Oosterlee [3], Recchioni and Sun [12] extended the model of Grzelak and Oosterlee [3] to a Heston

S. Bayad · K. Hilal
LAMSC Laboratoty, Sultan Moulay Slimane University, Beni Mellal, Morocco

A. El Hajaji (✉)
LESJEP Laboratory, FSJESJ, University Chouaib Doukkali, El jadida, Morocco
e-mail: a_elhajaji@yahoo.fr

S. Melliani et al. (eds.), *Applied Mathematics and Modelling in Finance, Marketing and Economics*, Studies in Computational Intelligence 1114,
https://doi.org/10.1007/978-3-031-42847-0_16

multi-factor model. He and Zhu [4] considered a Heston-CIR hybrid model where the interest rate follows the CIR model. They derived a closed-form pricing formula for European options in the form of an infinite series.

In this paper, we adopt the Heston-LS (Longstaff Schwartz [9]) hybrid model for the underlying price and we aim to present a closed-form pricing formula for European options as models with exact and analytical solutions.

The rest of the paper is organized as follows. In Sect. 2, a brief introduction of the Heston-LS model is given. In Sect. 3, a closed-form formula for European options is obtained with analytical solution. In Sect. 4, some numerical illustrations are given by computing European call option prices. Subsequently, some concluding remarks were made in Sect. 5.

2 The Dynamics of the Heston-LS Model

In this section, we will highlight European pricing option by using the Heston-LS model. This model is a hybrid model incorporating the Heston Stochastic Volatility Model with a two-factor interest rate model. The short-term interest rate and the uncertainty of the short-term interest rate outlined by Longstaff and Shwartz [9] are the two factors. The model dynamics under the a risk-neutral measure $\mathbb{Q}$ are specified as follows:

$$\begin{cases} \frac{dS_t}{S_t} = (\alpha \mathrm{x} + \beta \mathrm{y})\mathrm{dt} + \sqrt{v_t} dW_{s,t} \\ dv_t = k(\theta - v_t)dt + \sigma\sqrt{v_t} dW_{v,t} \\ dx_t = (\gamma - \delta x)dt + \sqrt{x_t} dW_{x,t} \\ dy_t = (\eta - sy)dt + \sqrt{y} dB_{y,t}, \end{cases} \tag{1}$$

where S_t is the underlying price, v_t is the volatility, x_t and y_t are state variables, $\gamma,\ \delta,\ \eta$, and s are constants. The expected value taken under the risk-neutral measure $\mathbb{Q}$ is denoted by E and assume the following structure of correlation:

$$\mathrm{E}(dW_{s_t} dW_{v_t}) = \rho dt, \quad t > 0, \quad \rho \in (-1, 1) \tag{2}$$

$$\mathrm{E}(dW_{x_t} dWs_t) = \mathrm{E}(dW_{x_t} dW_{v_t}) = \mathrm{E}(dW_{x_t} dW_{y_t}) = 0, \quad t > 0 \tag{3}$$

$$\mathrm{E}(dW_{y_t} dW_{s_t}) = \mathrm{E}(dW_{y_t} dW_{v_t}) = 0, \quad t > 0. \tag{4}$$

The relationship between the pairs (x, y) and (r, V) determined by the relations:

$$r_t = \alpha x(t) + \beta y(t) \tag{5}$$

$$V_t = \alpha^2 x(t) + \beta^2 y(t), \tag{6}$$

where r_t is the short-term interest rate, V_t is the volatility of the short-term interest rate, α and β are nonnegative constants.

If we introduce a forward measure $\mathbb{Q}^T$ such that the price of a European call option $\mathrm{U}(S, v, x, y)$ can be expressed as follows:

$$\mathrm{U}(S, v, x, y) = P(x, y, t, T)\mathrm{E}^T[max(S_T - K), 0) \mid S_t] \tag{7}$$

where K representing the strike price, $P(x, y, t, T)$ represente the price of a T-maturity zero coupon bond under $\mathbb{Q}$.

Then by using the technique of the numeraire change and the formula of $P(x, y, t, T)$ (see Appendix), we can obtain that the model dynamic under the forward measure $\mathbb{Q}^T$ can be expressed as follows:

$$\begin{cases} \frac{dS_t}{S_t} = (\alpha \mathrm{x} + \beta \mathrm{y})\mathrm{dt} + \sqrt{v_t} dB_{s,t} \\ dv_t = k(\theta - v_t)dt + \sigma\sqrt{v_t} dB_{v,t} \\ dx_t = (\gamma - (\delta - E_2)x)dt + \sqrt{x_t} dB_{x,t} \\ dy_t = (\eta - (s - E_3)y)dt + \sqrt{y} dB_{y,t} \end{cases} \tag{8}$$

where

$$\begin{aligned} E_2 &= E_2(t; T) = (\delta - m)(1 - A(t; T)) \\ E_3 &= E_3(t; T) = (s - p)(1 - B(t; T)) \\ A(t; T) &= \frac{2m}{(\delta + m)e^{(m(T-t)-1)} + 2m} \\ B(t; T) &= \frac{2p}{(s + p)e^{(p(T-t)-1)} + 2p} \\ m &= \sqrt{2\alpha + \delta^2},\ p = \sqrt{2\beta + s^2},\ \xi = \gamma(\delta + m) + \eta(s + p) \end{aligned}$$

We are now prepared to price European options after the launch of the adopted Heston-LS hybrid model, which is Next section's key content.

3 Pricing Formula for European Options

In the context of Model (8), we derive a closed-form pricing formula for the European option, and in order to achieve it we use the same approach as Heston [5], He and Zhu [4] and by using the Maple [11] mathematical package, we will derive an analytical solution formula for the characteristic function of the underlying price. After taking $l = ln(S)$ and letting $f(y))$ denote the probability density function. Of the underlying price, we have to rewrite the Eq. (7) as:

$$\mathrm{U}(l, v, x, y) = P(x, y, t, T)[P_1 - KP_2] \tag{9}$$

where $P_1 = \int_{ln(K)}^{+\infty} e^y f(y)dy$, and $P_2 = \int_{ln(K)}^{+\infty} f(y)dy$.

In other words, we assume that the characteristic function of the underlying price is $F(\phi; t, T, l, v, x, y)$. The results of $F(\phi; t, T, l, v, x, y)$ are given by the next theorem.

Theorem 1 *If the underlying asset price follows the dynamic specified in Eq. (8) the characteristic function of the underlying asset price $F(\phi; t, T, l, v, x, y)$ can be derived as:*

$$F(\phi; t, T, l, v, x, y) = e^{C(\tau;\phi)+D(\tau,\phi)v+E(\tau,\phi)x+G(\tau,\phi)y+i\phi l} \tag{10}$$

where $\tau = T - t$, $C(\tau, \phi)$, $D(\tau, \phi)$, $E(\tau, \phi)$ and $G(\tau, \phi)$ is given by Eqs. (16)–(18), respectively.

Proof The definition of $F(\phi; t, T, l, v, x, y)$ is:

$$F(\phi; t, T, l, v, x, y) = \mathrm{E}^{\mathbb{Q}^T}[e^{i\phi lT} \mid l_t, v_t, x_t, y_t] \tag{11}$$

After applying the Feynman-Kac theorem, we can easily find that $F(\phi; t, T, l, v, x, y)$ should satisfy the following PDE

$$\frac{\partial F}{\partial \tau} = \frac{\partial F}{\partial l}(\alpha x + \beta y - \frac{1}{2}v) + \frac{1}{2}v\frac{\partial^2 F}{\partial l^2} + k(\theta - v)\frac{\partial F}{\partial v} + \frac{1}{2}\sigma^2 v\frac{\partial^2 F}{\partial v^2} + \sigma v \rho\frac{\partial^2 F}{\partial l \partial v}$$
$$+ (\gamma - (\delta + E_2)x)\frac{\partial F}{\partial x} + \frac{1}{2}\frac{\partial^2 F}{\partial x^2} + (\eta - (s + E_3))\frac{\partial F}{\partial y} + \frac{1}{2}\frac{\partial^2 F}{\partial y^2} \tag{12}$$

with the initial condition as

$$F \mid_{\tau=0} = e^{i\phi l} \tag{13}$$

If we assume that $F(\phi; t, T, l, v, x, y)$ takes the form of

$$F(\phi; t, T, l, v, x, y) = e^{C(\tau;\phi)+D(\tau,\phi)v+E(\tau,\phi)x+G(\tau,\phi)y+i\phi l} \tag{14}$$

and substitute into Eq. (12), we can obtain

$$\frac{dD}{d\tau} = \frac{1}{2}\sigma^2 D^2 + (i\phi\rho\sigma - k)D - \frac{1}{2}(i\phi + \phi^2) \tag{15}$$

$$\frac{dE}{d\tau} = \frac{1}{2}E^2 - (\delta - E_2)F + i\phi\alpha \tag{16}$$

$$\frac{dG}{d\tau} = \frac{1}{2}E^2 - (s - E_3)F + i\phi\beta \tag{17}$$

$$\frac{dC}{d\tau} = k\theta D + \gamma E + \eta G \tag{18}$$

with

$$D(0) = E(0) = G(0) = C(0) \tag{19}$$

The ODE which governs $D(\tau, \phi)$ is clearly actually a Riccati equation with constant coefficients,this implies that it can be solved easily with some basic algebraic calculations. On the other hand, since ODE that govern $E(\tau, \phi)$ and $G(\tau, \phi)$ are again a Riccati equation, the expression of $E(\tau, \phi)$ and $G(\tau, \phi)$ are very difficult to decide as there is a time-dependent coefficient. Using the mathematical package Maple, we can conveniently find an exact analytic solution of ODE (16) and ODE (17).

$$D = \frac{d - (\rho\sigma i\phi - k)}{\sigma^2} \cdot \frac{1 - e^{d\tau}}{1 - g e^{d\tau}} \tag{20}$$

where $d = \sqrt{(\rho\sigma i\phi - k)^2 + \sigma^2(i\phi + \phi^2)}$ and $g = \frac{(\rho\sigma i\phi - k) - d}{(\rho\sigma i\phi - k) + d}$

$$E = \frac{(m - \delta)[k_1 e^{q_1\tau} + k_2 e^{q_2\tau} + k_3 e^{q_3\tau} + k_4 e^{q_4\tau}]}{\lambda_1 e^{q_1\tau} + \lambda_2 e^{q_2\tau} + \lambda_3 e^{q_3\tau} + \lambda_4 e^{q_4\tau}} \tag{21}$$

$$G = \frac{(p - s)[H_1 e^{z_1\tau} + H_2 e^{z_2\tau} + H_3 e^{z_3\tau} + H_4 e^{z_4\tau}]}{\chi_1 e^{z_1\tau} + \chi_2 e^{z_2\tau} + \chi_3 e^{z_3\tau} + \chi_4 e^{z_4\tau}} \tag{22}$$

$$\begin{aligned} C = {} & \frac{k\theta}{\sigma^2}\{[d - \rho\sigma i\phi - k)]\tau - 2log(\frac{-1 + g e^{d\tau}}{g - 1}\}) \\ & + \gamma\{w_1\tau + \frac{2}{\alpha}log[\frac{c_1}{m}\frac{[(\delta - m)e^{-\frac{\tau}{2}(c_1 - m)} - (\delta + m)e^{q_4\tau}]}{(\delta - c_1)e^{-\frac{\tau}{2}}(c_1 - m) - (\alpha + m)e^{q_4\tau}}]\} \end{aligned} \tag{23}$$

where

$$\begin{aligned} q_1 &= \frac{(c_1 + 3m)}{2}, q_2 = \frac{(c_1 + m)}{2}, q_3 = \frac{-(c_1 - m)}{2}. q_4 = \frac{-(c_1 - 3m)}{2} \\ z_1 &= \frac{(c_1 + 3m)}{2}, z_2 = \frac{(c_1 + m)}{2}, z_3 = \frac{-(c_1 - m)}{2}. z_4 = \frac{-(c_1 - 3m)}{2} \\ \lambda_1 &= -\alpha(c_1 + \delta),\ \lambda_2 = (c_1 + \delta)(-\alpha - \delta^2 + \delta m), \lambda_3 = (c_1 - \delta)(-\alpha - \delta^2 + \delta m), \\ \lambda_4 &= -\alpha(c_1 + \delta), k_1 = [m - c_1 - a_7(m + \delta)], k_2 = [m + c_1 - a_7(m - \delta)] \\ k_3 &= [-m + c_1 + a_7(m - \delta)], k_4 = [-m + c_1 + a_7(m + \delta)], c_1 = \sqrt{(-2a_7\alpha + 2\alpha + \delta^2)} \\ \chi_1 &= -\beta(c_2 + s),\ \chi_2 = (c_2 + s)(-\beta - s^2 + sp), \chi_3 = (c_2 - s)(-\beta - s^2 + sp), \\ \lambda_4 &= -\beta(c_2 + s), H_1 = [p - c_2 - a_7(p + s)], H_2 = [p + c_2 - a_7(p - s)] \\ H_3 &= [-p + c_1 + a_7(p - s)], H_4 = [-p + c_1 + a_7(p + s)], c_2 = \sqrt{(-2a_7\beta + 2\beta + s^2)} \end{aligned}$$

$$d = \sqrt{(a_7\rho\sigma - k)^2 + \sigma^2(a_7 + \phi^2)},\ g = \frac{(a_7\rho\sigma - k) - d}{(a_7\rho\sigma - k) + d}$$

$$w_1 = \frac{2[a_7(\delta - m) + (m - c_1)]}{(\delta - c_1)(\delta - m)} + \frac{2(m - c_1)}{(\alpha^2 - m^2)} + \frac{6(m - c_1)(a_7 - 1)}{\delta^2 - c_1^2}$$

$$w_2 = \frac{2[a_7(s - p) + (p - c_2)]}{(s - c_2)(s - p)} + \frac{2(p - c_2)}{(s^2 - p^2)} + \frac{6(p - c_2)(a_7 - 1)}{s^2 - c_2^2}$$

$$m = \sqrt{(2\alpha + \delta^2)},\ p = \sqrt{(2\beta + s^2)} and a_7 = i\phi.$$

Moreover, from Heston [5], He and Zhu [4] and the formula of $F(\phi; t, T, l, v, x, y)$, we can obtain.

Theorem 2 *If the underlying asset price follows the dynamic specified in Eq. (8), the pricing formula for European call options is given by*

$$\mathrm{U}(l, v, x, y) = P(x, y, t, T)[P_1 - K P_2] \tag{24}$$

where

$$P_1 = F(-i; t, T, l, v, x, y)\{\frac{1}{2} + \frac{1}{\pi}\int_0^{+\infty} Re[\frac{e^{-i\phi lnK F(\phi - i; t, T, l, v, x, y)}}{i\phi F(-i; t, T, l, v, x, y)}]d\phi\} \tag{25}$$

$$P_2 = \frac{1}{2} + \frac{1}{\pi}\int_0^{+\infty} Re[\frac{e^{-i\phi lnK F(\phi; t, T, l, v, x, y)}}{i\phi}]d\phi \tag{26}$$

4 Numerical Experiments

in this section, we present some numerical results for the verification of European call option prices under Heston-LS hybrid model parameters. In the following, unless otherwise stated, parameters we use are listed as follows.The mean-reverting speed $k = 10$, the long-term mean θ and the volatility of volatility σ take the value of 10, 0.2 and 0.1 respectively,while the corresponding parameters for LS model satisfy $\alpha = 0.005$, $\beta = 0.0814$, $\eta = 3.2033$, $s = 14.4227$, $\gamma = 4.0224$, $\delta = 0.3299$, $V_0 = 0.00081$ and $r_0 = 0.06717$. The strike price $K = 100$, the underlying priceS $S_0 = 100$, and the time to expiry τ is 1.

In the pricing of options, the strike price plays a significant role. One of the biggest challenges in the finance sector is deciding a realistic strike price for options. European call option prices are obtained by considering the various strike price values of K, $r0$, and V_0 (see Tables 1 and 2). The findings obtained show that a rise in the value of the strike price is leading to a decrease in the value of the European call option.

Moreover, both the initial value of underlying price and expiration date play an important role in option pricing. Here, we investigate the value of the European call option by considering different values of the s_0 and expiration date (Tables 3 and 4).

Table 1 European call option price with respect to different values of the strike price K and r_0

K	$r_0 = 0.040000$	$r_0 = 0.057$	$r_0 = 0.06$	$r_0 = 0.067$
90	24.3913	25.0757	25.1971	25.1682
94	22.3323	22.9966	23.1145	23.0865
98	20.4238	21.0656	21.1796	21.1526
100	19.5242	20.1539	20.2659	20.2393

Table 2 European call option price with respect to different values of the strike price K and V_0

K	$V_0 = 0.0007$	$V_0 = 0.00075$	$V_0 = 0.00081$	$V_0 = 0.00086$
90	25.4685	25.4774	25.1682	25.4969
94	23.3785	23.3871	23.0865	23.4056
98	21.4351	21.4434	21.1526	21.4617
100	20.5167	20.5249	20.2393	20.5428

Table 3 European call option price with respect to different values of the initial value of underlying asset price s_0 and r_0

S_0	$r_0 = 0.04$	$r_0 = 0.05700$	$r_0 = 0.06$	$r_0 = 0.067$
90	14.2727	14.1302	14.2197	14.1985
94	16.5820	16.425	16.5236	16.5002
98	19.0457	18.8743	18.9820	18.9564
100	20.3322	20.1539	20.26594	20.2393

Table 4 European call option price with respect to different values of the initial value of underlying asset price S_0 and V_0

S_0	$V_0 = 0.0007$	$V_0 = 0.00075$	$V_0 = 0.00081$	$V_0 = 0.00086$
90	14.4204	14.427	14.1985	14.4413
94	16.7448	16.751971	16.5002	16.7678
98	19.2230	19.230931595	18.9564	19.2481
100	20.516	20.524942	20.2393	20.542889

Obtained results reveal that the increase in the value of the S_0 result in increase the value of European call option.

On the other hand. Figures 1 and 2 show Option prices under our model, CIR model and the Heston model with respect to the underlying price and time to expiry It can be clearly seen that our price under the selected set of parameters is higher than the Heston price.

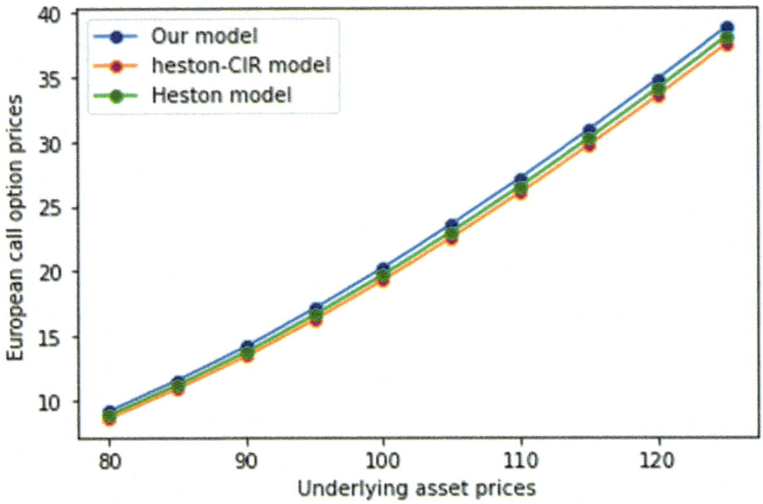

Fig. 1 Our price, Heston-CIR price and the Heston price with respect to the underlying asset price

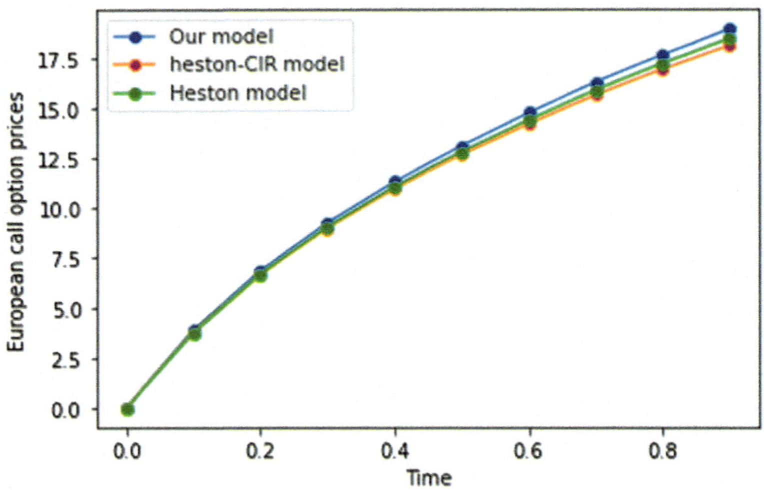

Fig. 2 Our price, Heston-CIR price and the Heston price with respect to the time to expiry

5 Appendix

If the risk-free interest rate follows the Longstaff Schwartz model given in Eq. (5), then we can easily find that the price of a T-maturity zero coupon bond $P(x, y, t, T)$ should satisfy the following PDE system:

$$\begin{cases} \frac{\partial P}{\partial t} + (\gamma - \delta x)\frac{\partial P}{\partial x} + (\eta - sy)\frac{\partial P}{\partial y} + \frac{x}{2}\frac{\partial^2 P}{\partial x^2} + \frac{y}{2}\frac{\partial^2 P}{\partial x^2} - (\alpha x + \beta y)P = 0 \\ P(x, y, T, T) = 1 \end{cases} \tag{27}$$

If we assume that $P(x, y, t, T)$ takes the form of

$$P(x, y, t, T) = E_1(t, T)e^{(E_2(t,T)x + E_3(t,T)y)} \tag{28}$$

and substitute it into PDE (27), we can obtain

$$\begin{cases} \frac{\partial E_2}{\partial t} = \frac{1}{2}E_2^2 - \delta E_2 + \alpha \\ \frac{\partial E_3}{\partial t} = \frac{1}{2}E_3^2 - sE_3 + \beta \\ \frac{\frac{\partial E_1}{\partial t}}{E_1} = \gamma E_2 + \eta E_3 \end{cases} \tag{29}$$

with the terminal condition $E_1(T, T) = 0$, $E_2(T, T) = 0$ and $E_3(T, T) = 0$. We have equations that governs $E_2(t, T)$ and $E_2(t, T)$ are actually a Riccati equation that, with some algebraic calculation, can be easily solved. The expression of $E_1(t, T)$ will then be obtained by direct integration. The proof has been completed.

6 Conclusion

In this paper, it is presumed that the price of the underlying asset follows the Heston stochastic volatility model, with interest adopting two-factor model of Longstaff and Schwartz, we present a closed-form pricing formula for European option with analytical solution. The numerical results show that European call option prices under the Heston-LS model is higher than that under the Heston model and Heston-CIR model.

References

1. Dumas, B., Fleming, J., Whaley, R.E.: Implied volatility functions: Empirical tests. J. Finance. **53**(6), 2059–2106 (1998)
2. Black, F., Scholes, M.: The pricing of option and corporate liabilities. J. Polit. Econ. **81**(3), 637–59 (1973). https://doi.org/10.1086/260062
3. Grzelak, L.A., Oosterlee, C.W.: On the Heston model with stochastic interest rates. SIAM J. Financ. Math. **2**(1), 255–86 (2011). https://doi.org/10.1137/090756119
4. He, X.J., Zhu, S.P.: A closed-form pricing formula for European options under the Heston model with stochastic interest rate. J. Comput. Appl. Math. **335**, 323–33 (2018). https://doi.org/10.1016/j.cam.2017.12.011
5. Heston, S.L.: A closed-form solution for options with stochastic volatility with applications to bond and currency options. Rev. Financ. Stud. **6**(2), 327–343 (1993). https://doi.org/10.1093/rfs/6.2.327
6. Johnson, H., Shanno, D.: Option pricing when the variance is changing. J. Financ. Quant. Anal. **22**(02), 143–151 (1987)
7. Wiggins, J.B.: Option values under stochastic volatility: theory and empirical estimates. J. Financ. Econ. **19**(2), 351–372 (1987)
8. Hull, J., White, A.: The pricing of options on assets with stochastic volatilities. J. Finance. **42**(2), 281–300 (1987)
9. Longstaff, F.A., Schwartz, E.S.: Interest-rate volatility and the term structure: a two-factor general equilibri-um model. J. Financ. Forthcom. (1992)
10. Scott, L.O.: Option pricing when the variance changes randomly: theory, estimation, and an application. J. Financ. Quant. Anal. **22**(04), 419–438 (1987)
11. Maplesoft, Maple: 12 Users Manual. Maplesoft, Waterloo (2008)
12. Recchioni, M.C., Sun, Y.: An explicitly solvable Heston model with stochastic interest rate. Eur. J. Oper. Res. (2015). https://doi.org/10.1016/j.ejor.2015.09.035

A Multi-objective Approach to Energy Efficiency in Cellular Networks

Soufiane Dahmani and Abdelhafid Serghini

Abstract In the current generation of cellular networks, energy efficiency is considered as an important issue due to their high-energy consumption. To meet the rising traffic resulting from the growing mobile stations requests and covering the entire transmission area, base stations must be increasingly deployed. However, increasing the number of these stations increases the cost and energy consumption, which leads to conflicting goals. In this paper, we introduce a multi-objective mathematical model on cellular networks that aimed to minimize the expected total cost of base stations and maximize total coverage. This optimization must take into account the traffic demand profile. Given that the studied model corresponds to an NP-hard multi-objective problem, we use a meta-heuristic algorithm to solve it. The simulation results show the effectiveness of our approach to cover the grid while reducing costs and energy consumption.

1 Introduction

The wireless world that provides users with high speed internet use and full coverage has undoubtedly increased the demand for traffic, spurred the equitable development of new infrastructure and rapid growth in energy demand. This leads to an increased demand for natural energy that accounts, as an example, for more than 18% of operational cost In European countries. In front of this situation, more attention is being focused on the optimization of energy consumption costs.

A basic requirement for less expensive wireless connections is low power consumption. However, the increase in the number of mobile stations (MS) and users needs drive the mobile operators to increment the number of base stations (BS), which

S. Dahmani (✉) · A. Serghini
ANAA Research Team, ESTO, LANO Laboratory, FSO, University Mohammed First, 60050 Oujda, Morocco
e-mail: s.dahmani@ump.ac.ma

A. Serghini
e-mail: a.serghini@ump.ac.ma

S. Melliani et al. (eds.), *Applied Mathematics and Modelling in Finance, Marketing and Economics*, Studies in Computational Intelligence 1114,
https://doi.org/10.1007/978-3-031-42847-0_17

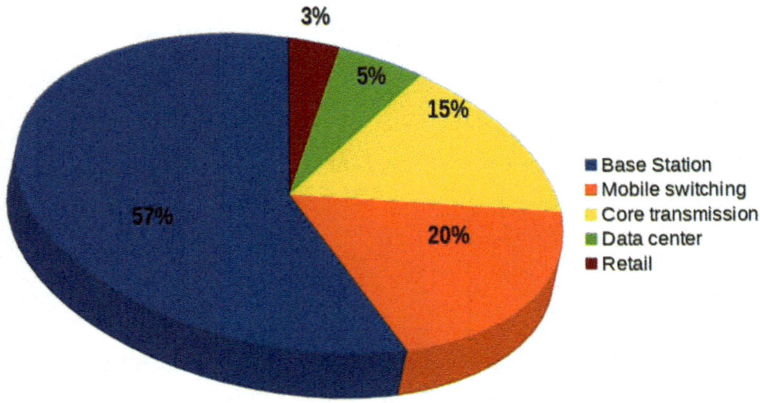

Fig. 1 Power consumption of wireless cellular network

increases automatically the power consumption. As a consequence, the expected total cost of the cellular network increased also. Figure 1 shows that BS is the main energy consumer of the wireless cellular network [1].

Despite all the attention paid to grid energy consumption, recent data shows that the amount of energy used by the grid continues to increase. Against this backdrop, the main challenge of the present work is to, (i) minimize the total cost of the base stations which would minimize the energy consumption without compromising the quality of service (QoS). (ii) maximize network coverage to enhance service quality.

In particular, in this article, that refines our previous work [2–4], we focus on improving the total cost given by the sum of the installation cost and the usage cost of the electrical system that power base stations. In this optimization, We take into account the service quality and the expected changes in conditions that include the strong growth in traffic demand from end-users, as well as the increase in the price of the electricity network. Also, we consider the evolution of base station costs and the increase in installation costs from year to year.

This document is organized as follows. Section 2 presents previous work in the field. Section 3 formulates the problem of optimizing the total cost of base stations. Section 4 proposes a meta-heuristic algorithm to solve the problem. Section 5 presents the results of the simulation and Section 6 concludes the article.

2 Related Work

Many studies and research have been conducted on energy efficiency for BSs in cellular networks. Some of the earliest works only addressed the reduction of the base stations' power consumption. Others dealt with enhancing the service quality for users and look to guarantee the full coverage when they decide the status of the BS [2, 3].

BS traffic load is the common decision metric in most previous research. In [2, 5], a traffic sensitive algorithm has been proposed to enable and disable BS LTE. In [6], the amount of energy reduction in the network is studied by reducing the number and size of active macrocells as a function of the traffic load. In this article, BS's sleep pattern is determined based on two aspects. The first is the ability to provide coverage from neighboring BSs, and the second is to disable the maximum possible number of BSs to ensure that power consumption is minimized.

Looking at the details of an energy system, although there have been several published works in the field of energy efficiency (e.g. [7, 8]), only a few articles have appeared in the telecommunications context. The main purpose of these articles is to optimize the energy consumption of the base station while considering the variation of the energy consumed for a variable traffic load. In line with this idea, [9] proposed optimization of the cost generated by a hybrid system used to power a BS GSM/CDMA. This optimization takes into account the annual variations of wind and solar energy without considering the daily traffic variation.

Several works have focused on the allocation of resources in a part of a network comprising a few BS. For example, [10] introduced an optimal BS on/off strategies for saving power within a radio access network. The same authors investigated a problem of association of green energy conscious users and latency in [11]. Dahmani et al. [2] considered a similar problem to the one addressed by 11 but more easier as they did not consider in detail how the energy is consumed and the total coverage of the users. He also proposed an on-off strategy that does not take into account the cost and the place of installation related to each BS. This restriction leads to ignoring the variation in energy production.

From the above review, we conclude that most of the previous base station power reduction methods did not look at the overall coverage after the power consumption optimization procedure. They did not also interested in the total cost of these base stations. While some studies attempted to solve the common problem of shutting down BS and covering most users, they focued on system energy efficiency where coverage is as important as energy efficiency and should be.

3 System Model and Problem Formulation

In this section, we formally present the studied problem and describe the mathematical formulation that we will study. We consider a territory covered by cellular network service for our work.

3.1 *Traffic State of the Mobile Users*

Having a detailed daily definition of the traffic generated by mobile users is essential for calculating the daily power consumption of the base station as well as knowing the number of stations that will be in operation for the total coverage of these users, and the stations that will be shut down. Suppose that in a given cellular network coverage area and for a day, the traffic is distributed as follows:

- Between 00:00 and 12:00, the traffic does not exceed 600 GB;
- Between 12 a.m. and 2 p.m., traffic is between 600 and 900 GB;
- Between 2 p.m. and 8 p.m., traffic exceeds 900 GB;
- Between 8 p.m. and 10 p.m., traffic is between 600 and 900 GB;

From this data, the traffic variations can be divided into three different traffic states s_1, s_2 and s_3 or:

- $s1 = [0; 600]$,
- $s2 = [600; 900]$
- $s3 = [900; 1200]$.

Figure 2 illustrates these traffic values and their variation during a day.

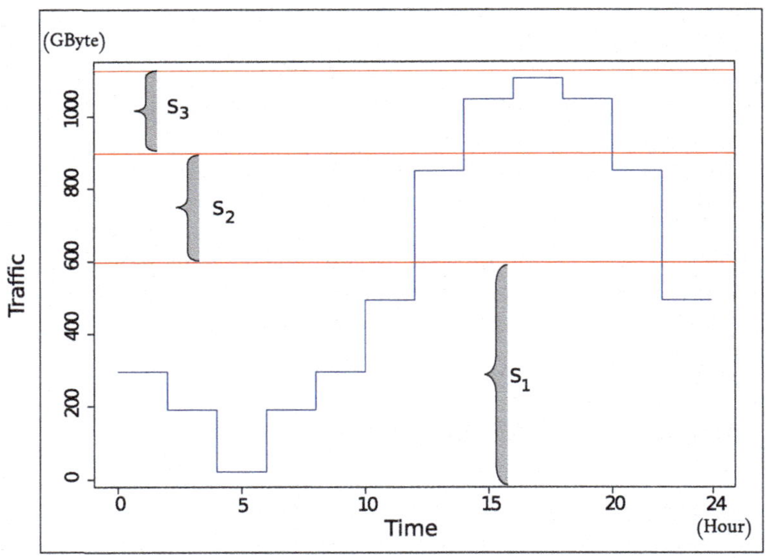

Fig. 2 The mobile stations' traffic over a given two-hour time is the monthly average of traffic randomly generated across the same time interval

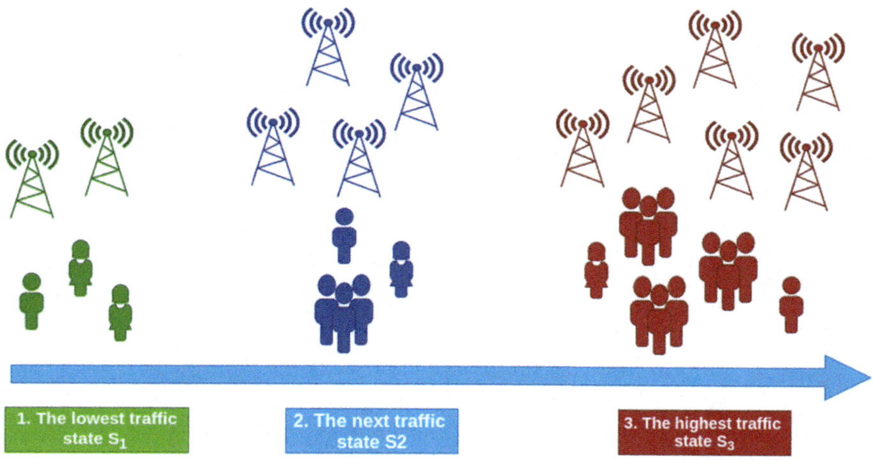

Fig. 3 Approach proactive

3.2 *Approach Proactive*

In this approach, we choose from $\mathscr{B}$ the optimal set of BS_s at the lowest traffic state. This set of BS_s can be used at any time, in the network, for all traffic states. Then, additional BS_s are turned on from a traffic state to another as the traffic increases to meet the increasing capacity and coverage requirements. In Fig. 3, if we use this approach, we must start by the first traffic state s_1 to the last one which is s_3, for more details see [2, 3].

3.3 *Problem Modeling*

We consider a service area that we want to cover by a cellular network service. Let $S = \{1, ..., m\}$ a set of base stations (BS_s) which will be installed and $I = \{1, ..., n\}$ a set of mobile stations (MS_s). Each base station BS_i has an installation cost denoted c_i. We denote by u_j the number of connections simultaneously active by the mobile station MS_j. Consider the following two decision variables:

$$x_i = \begin{cases} 1 \text{ if } BS_i \text{ is selected,} \\ 0 \text{ otherwise.} \end{cases} \tag{1}$$

$$y_{k,i} = \begin{cases} 1 \; if \;\; MS_k \;\; is \;\; served \;\; by \;\; BS_i, \\ 0 \;\; \text{otherwise.} \end{cases} \tag{2}$$

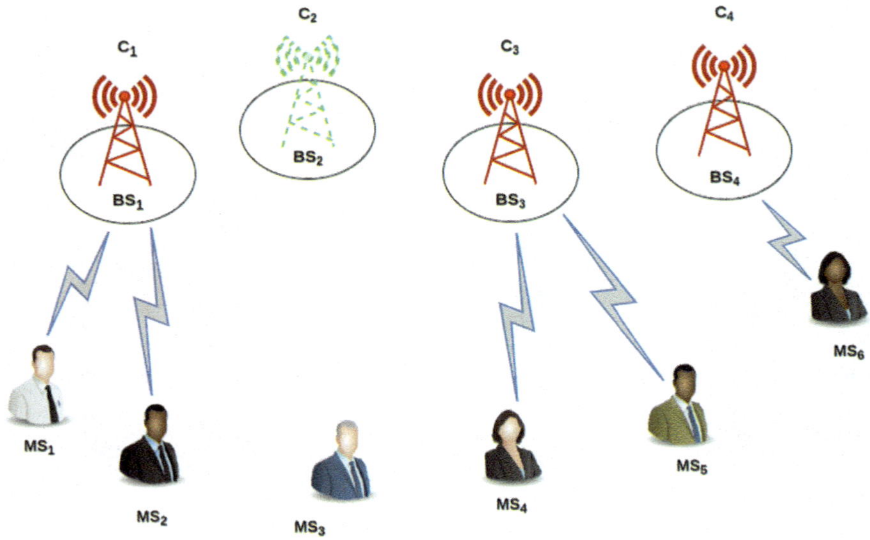

Fig. 4 Illustration: problem with six mobile stations and four base stations

An illustrative example is shown in Fig. 4. In this example, we have four base stations (BS_s) and six mobile stations (MS_s). We notice that the BS_2 is not installed and the MS_3 is not covered; MS_1 and MS_2 are assigned to BS_1; MS_4 and MS_5 are assigned to BS_3 and MS_6 is assigned to BS_4.

Since we want to maximize the total covered traffic and minimize the total cost of base station under certain constraints, the problem can be expressed as follows:

$$\begin{cases} \min\limits_{x} \sum\limits_{i=1}^{N} c_i x_i \\ \max\limits_{y} \sum\limits_{s \in S} \sum\limits_{k=1}^{K_s} \sum\limits_{i=1}^{N} u_k y_{k,i} \end{cases} \tag{3}$$

Subject to the following constraints:

$$\begin{aligned} & x_i P_{k,i} - SINR_k \sum_{j=1, j \neq i}^{N} x_j P_{k,j} - SINR_k \sigma^2 \geq \\ & \left(-SINR_k \sum_{j=1, j \neq i}^{N} P_{k,j} - SINR_k \sigma^2 \right) (1 - y_{k,i}) \quad \forall k \in J, \forall i \in I, \end{aligned} \tag{4}$$

$$y_{k,i} \leqslant x_i \qquad \forall k \in J, \forall i \in I, \tag{5}$$

$$\sum_{i=1}^{N} y_{k,i} \leqslant 1 \qquad \forall k \in J, \tag{6}$$

$$\sum_{k=1}^{K_s} y_{k,i} \leqslant K_{BS} \qquad \forall i \in I, \tag{7}$$

$$y_{k,i} \in \{0, 1\}, x_i \in \{0, 1\} \qquad \forall k \in J, \forall i \in I. \tag{8}$$

Constraint (4) represents the quality of service of mobile stations (MS) in the network. The constraint (5) ensures that one cannot assign the mobile station number k (MS k) to be served by the base station number i (BS i) if BS i is not selected and is in service. Constraint (6) forces each MS k to be served by at most one BS. The constraint (7) guarantees that each BSi can serve at most K BS MS. Constraint (8) requires that the variables xi and yki be binary according to Eqs. (1) and (2).

3.3.1 Improved Formulation of the Problem

We transform this multi-objective problem into a mono-objective one using the weighted sum method as follows:

$$\max_{x,y} \alpha \sum_{s\in S} \sum_{k=1}^{K_s} \sum_{i=1}^{N} u_k y_{k,i} - (1-\alpha) \sum_{i=1}^{N} c_i x_i \tag{9}$$

subject to the constraints (4), (5), (6), (7) and (8).

Here, we use **dynamic weights** instead of constant ones.

$$\begin{cases} \max\limits_{x,y} \alpha(t) \sum\limits_{s\in S} \sum\limits_{k=1}^{K_s} \sum\limits_{i=1}^{N} u_k y_{k,i} - (1-\alpha(t)) \sum\limits_{i=1}^{N} c_i x_i \\ |\alpha(t) \sum\limits_{s\in S} \sum\limits_{k=1}^{K_s} \sum\limits_{i=1}^{N} u_k y_{k,i} + (1-\alpha(t)) \sum\limits_{i=1}^{N} c_i x_i| < \varepsilon \end{cases} \tag{10}$$

subject to the constraints (2) to (7), where:

ε is a positive number very close to 0,

t is a time-step,

$\alpha(t)$ and $(1-\alpha(t))$ are dynamic weights of the two objective functions.

4 Simulated Annealing Algorithm Solution

The problem (13) formulated previously is NP-hard, so it is necessary to propose an approximate algorithm to solve it. In this paper, we use a simulated annealing (SA) algorithm. Simulated annealing (SA) is a probabilistic metaheuristic, inspired by the natural annealing process used in metallurgy, proposed by [12].

4.1 The Neighborhood of a Solution

Consider "N" base stations (BS) and Ks mobile stations (MS). A solution y to our problem is a sequence of Ks digits, where each digit is an integer taking values between 0 and N. In this article, we define a neighborhood $\bar{y}$ of a solution y by changing a digit into y by another integer from the set 0, ..., N. Figure 5 illustrates this mutation to generate a neighborhood of our solution.

4.2 SA Algorithm Solution

The simulated annealing (SA) algorithm is illustrated by Algorithm 1.

Algorithm 1 SA

Initialize the initial temperature T_0 and the final temperature T_f.
Generate a random solution y_0
WHILE $(T_F < T_0)$
FOR (a predetermined number of times) **do**
Choose, randomly, a neighborhood $\bar{y}_0$ of y_0
Compute $\Delta E = f(\bar{y}_0) - f(y_0)$, where f is the function defined by Eq. 9.
If $\Delta E > 0$, then $\bar{y}_0$ is the new state.
Else $\bar{y}$ is the new state with the probability $e^{-\Delta E/T}$.
ENDIF
ENDFOR
Decrease the temperature.
ENDWHILE

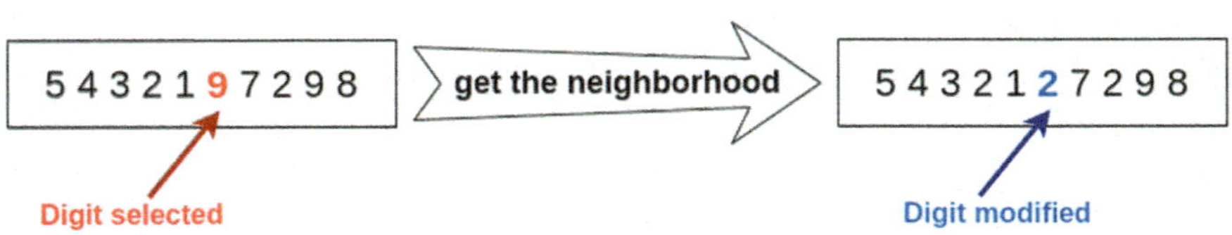

Fig. 5 Neighborhood operator

5 Simulation Results and Application

5.1 Description of Data

To evaluate the performance of the proposed algorithm, we consider a rectangular service area, a number of base stations and a number of mobile stations. With a pseudo-random number generator, each base station and each mobile station take a location in the service area according to a uniform distribution. For the general simulation parameters are:

- The maximum transmission power is equal to 20W,
- The power of thermal noise is equal to 5.97×10^{-15} W,
- The Carrier Frequency is equal to 2000 MHz,
- The maximum number of MS that can be served by each BS is equal to 50,
- The minimum threshold value that SINR is equal to −5dB,
- And the Gain of transmitter and receiver is equal to 0 dBi.

In order to evaluate the performance of our approach, we have considered two instances of our problem. We start by using these instances test to evaluate the capability of our approach to minimize the base stations' number, thus minimizing the total cost. Then, a proactive approach is introduced to minimize daily energy consumption generated by the installed BSs. The size of the service area is 10×10 (Km^2). We randomly generated the locations of 1200 mobile stations (MSs) and 150 base stations (BSs). the costs of the base stations are taken randomly between two digital units (Table 1).

5.2 Computational Results

This section reports the calculation results obtained by applying the proposed approach. Simulated annealing is encoded in C ++ language and executed on an Intel Core i5-4200U CPU 1.60GHz × 4. The parameters of SA are set as follows.

Table 2 shows the number of active base stations with a minimal cost total and MSs not covered. The results are obtained by simulated annealing (SA).

Table 1 Parameters of simulated annealing

Parameters	Values (s_1, s_2, s_3)
Initial temperatures	(100, 150, 200)
Iterations per temperature	(200, 250, 300)
Final temperature	0.001
Exponential cooling	0.09

Table 2 The result for the number of BS installed and MS not covered using classic and proactive approach

Our model	Traffic states	BS_s installed with minimal cost	MS_s not covered
Classic	Full day	140	1
Proactive	S_1	87	0
	S_2	120	1
	S_3	140	1

After having presented two possible solutions to reduce the energy consumption of the cellular network and thus reduce the total cost of the base stations, we conider the following example to illustrate the descrobed approach. If the value 'A' is the energy consumed by a base station BSi during one hour, then we can calculate the total energy consumed (Ec) by this station during the day, with the following term:

$$Ec = Ts_i \times NB_i \times A$$

where,

- NB_i is the number of BSs selected from the traffic state $s_i, i \in \{1, 2, 3\}$;
- Ts_i in hour is the duration of the traffic state $s_i, i \in \{1, 2, 3\}$;
- A is is the energy consumed by a base station BS_i during one hour = 40 W.

therefore for the classical method we obtain the following result:

$$Ec = 24 \times 140 \times 40 = 134,4 \times 10^3 \quad W/day$$

on the other hand in a proactive approach and on/off strategy we obtain:

$$Ec(s_1) = 14 \times 87 \times 40 = 48720 \quad W$$

$$Ec(s_2) = 4 \times 120 \times 40 = 19200 \quad W$$

$$Ec(s_3) = 6 \times 140 \times 40 = 33600 \quad W$$

$$Ec = Ec(s_1) + Ec(s_2) + Ec(s_3) = 101,52 \times 10^3 \quad W/day$$

We see that when we use the proactive approach, the energy consumption is decreased by **24.46 %**.

The results depicted above clearly show the ability of our approach to providing an optimal solution that minimizes the total cost while reducing the number of BSs and keeping the maximum coverage in the classical application. Another interesting result is the capacity of the proactive approach in improving the solution quality.

6 Conclusion

Energy efficiency is a growing concern, particularly in the wireless communication networks of today and tomorrow. This is due to the sharp increase in the number of users and the continued needs of these networks. Therefore, energy consumption is expected to increase significantly. This leads to serious energy and economic problems. In this article, we have studied this problem from this aspect and tried to reduce the total cost of base stations in the cellular network and thus reduce the energy consumption of these stations. The second challenge is to maximize the total coverage in order to maintain the quality of service. First we classified this problem as a multipurpose problem. Second, we proposed an on/off strategy based on the annealing simulation method to solve the NP-hard problem.

References

1. Han, C., Harrold, T., Armour, S., Krikidis, I., Videv, S., Grant, P.M., Haas, H., Thompson, J.S., Ku, I., Wang, C.-X., et al.: Green radio: radio techniques to enable energy-efficient wireless networks. IEEE Commun. Mag. **49**(6), 46–54 (2011)
2. Dahmani, S., Gabli, M., Mermri, E.B., Serghini, A.: Optimization of green RNP problem for LTE networks using possibility theory. Neural Comput. Appl. **32**(8), 3825–3838 (2020)
3. Mohammed, G., Soufiane, D., El Bekkaye, M., Abdelhafid, S.: Optimization of multi-objective and green LTE RNP problem. In: 2019 International Conference on Wireless Technologies, Embedded and Intelligent Systems (WITS), pp. 1–6. IEEE (2019)
4. Gabli, M., Dahmani, S., Mermri, E.B., Serghini, A.: A dynamic genetic algorithm approach to the problem of UMTS network assignment. In: International Conference on Networked Systems, pp. 15–26. Springer (2017)
5. Saxena, N., Sahu, B.J., Han, Y.S.: Traffic-aware energy optimization in green LTE cellular systems. IEEE Commun. Lett. **18**(1), 38–41 (2013)
6. Alsharif, M.H., Nordin, R., Ismail, M.: Cooperation management among base stations based on cells switch-off for a green LTE cellular network. Wirel. Pers. Commun. **81**(1), 303–318 (2015)
7. Merei, G., Berger, C., Sauer, D.U.: Optimization of an off-grid hybrid PV-Wind-Diesel system with different battery technologies using genetic algorithm. Sol. Energy. **97**, 460–473 (2013)
8. Sinha, S., Chandel, S.: Review of software tools for hybrid renewable energy systems. Renew. Sustain. Energy Rev. **32**, 192–205 (2014)
9. Nema, P., Nema, R., Rangnekar, S.: PV-solar/wind hybrid energy system for GSM/CDMA type mobile telephony base station. Int. J. Energy Environ. **1**(2), 359–366 (2010)
10. Gong, J., Thompson, J.S., Zhou, S., Niu, Z.: Base station sleeping and resource allocation in renewable energy powered cellular networks. IEEE Trans. Commun. **62**(11), 3801–3813 (2014)
11. Han, T., Ansari, N.: Green-energy aware and latency aware user associations in heterogeneous cellular networks. In: 2013 IEEE Global Communications Conference (GLOBECOM), pp. 4946–4951. IEEE (2013)
12. Kirkpatrick, S., Gelatt, C.D., Vecchi, M.P.: Optimization by simulated annealing. Sci. **220**(4598), 671–680 (1983)

Inverse Problem of 2D Lung Electrical Impedance Tomography

Soumaya Idaamar and Mohamed Louzar

Abstract Electrical impedance tomography is a technique that allows to image the distribution of the conductivity of a domain from impedance measurements made at several points on its surface. The method was initially developed by geophysicists for mineral prospecting (Maillet, 1947). However, its biomedical applications were quickly recognized (Brown and Barber, 1984). Electrical Impedance Tomography (EIT) is a non-invasive imaging technology that estimates the electrical conductivity distribution in a domain. In this study, the conductivity is reconstructed from boundary voltage measurements by using a reconstructing algorithm known as the forward problem. Simultaneously, the image reconstruction can be obtained using the inverse problem to detect regional lung ventilation.

1 Introduction

Medical imaging allows to image the internal structure of the human body, it takes into account tissue properties such as the conductivity or the permittivity. Currently, some applications are in clinical use for the diseases diagnosis such as gastric emptying monitoring, regional lung surveillance, cardiac function and breast cancer.

Today, new technologies have made it possible to develop non-invasive monitoring tools available at the patient's bedside. Some of these tools could allow individual and more precise adaptation of the recommendations to adjust the tidal volume and positive expiratory pressure. A few examples of technologies for measuring volumes and pressures, or even new pulmonary imaging modalities is the electrical impedance tomography(EIT) [1].

S. Idaamar (✉) · M. Louzar
Department of Computer mathematics and engineering sciences, University Hassan I FST Settat, Km 3, B.P: 577 Road of Casablanca, Settat, Morocco
e-mail: s.idaamar@uhp.ac.ma

M. Louzar
e-mail: mohamed.louzar@uhp.ac.ma

S. Melliani et al. (eds.), *Applied Mathematics and Modelling in Finance, Marketing and Economics*, Studies in Computational Intelligence 1114,
https://doi.org/10.1007/978-3-031-42847-0_18

The development in medical EIT is motivated by its low cost, its non-ionizing character and its simplicity of use with a portable measurement system [2].

EIT applications to monitoring regional lung function are based on bioimpedance measurements by using a belt of electrodes containing 16 electrodes placed around the chest wall and an alternated current injected into a pair of adjacent electrodes which produced 13 pieces of measured voltage data. However, no voltage is measured between the electrodes where the current is injected [3].

The study aimed to determine the electrical properties of body tissues. The first step is to solve the forward problem using the finite element method to estimate potentials inside and at the boundaries of a domain, then to reconstruct the electrical conductivity distributions the inverse problem method was used, the measurement data are processed by the MATLAB program using the Gauss-Newton regularized iterative method [4].

2 Mechanism of Operation

Electrical Impedance Tomography (EIT) is a technique for imaging the electrical conductivity distribution of a section of the body based on a set of potential measurements on the surface.

The physical substance of the human body organism is composed of a wide variety of biological tissues with very different properties, it can be considered as a composite linear, homogeneous conductor. Thus, the electric resistivity of human tissues have a different value [5].

For the pulmonary system, the model had five regions: lungs, heart, blood, bone and fat, a conductivity was assigned to each of these regions using typical values obtained from the literature [6]. Tissue resistivity values is mentioned in the following Table 1.

The process of monitoring regional lung can be divided into two distinct phases: The first phase is the determination of the potential distribution in the domain. This method is known as 'The forward problem' with an initial given conductivity (Fig. 1). The second phase or 'The inverse problem' is the reconstruction of con-

Table 1 Values tissue resistivity

Tissues	Resistivity values (Ω_m)
Blood	1.6
Heart	2.5
Lungs	20
Bone	177
Fat	25

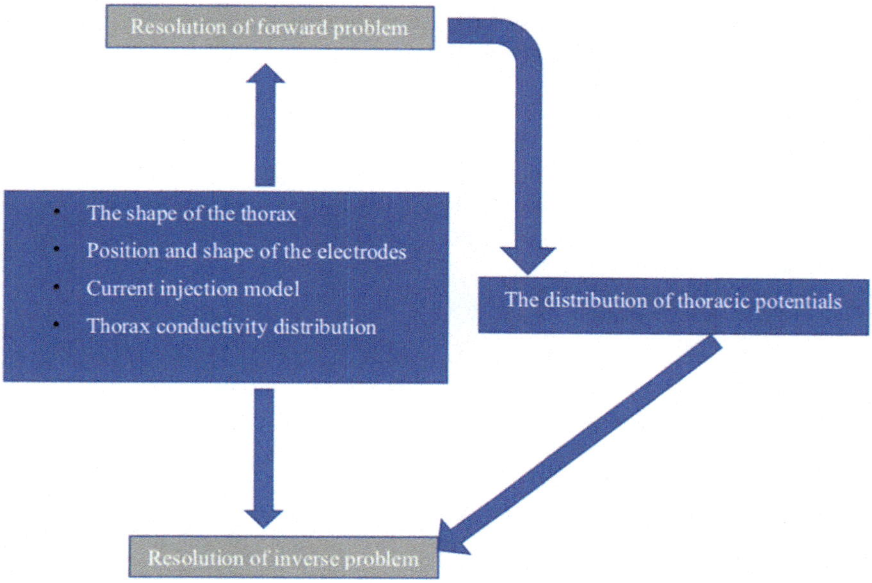

Fig. 1 Definition of forward and inverse problem

ductivity distribution. The forward and inverse problem are defined in the following diagram [7]

3 The Forward Problem

The forward problem can be solved analytically for simple forms by Fourier transforms [8].

For complex geometries, it is necessary to use numerical methods to discretize the domain into small elements, the method used is the Finite Element Method (FEM), it is generally chosen for EIT applications since it does not require any discretization regularity. The variational method is widely defined in the literature [9].

3.1 The Mathematical Model of the Forward Problem

The mathematical model of the studied problem are a simplification of Maxwell's equations as well as the related auxiliary relations and hypothesis of frequencies to obtain the Laplace equation with boundary conditions defined by

$$\begin{cases} -\nabla.(\sigma\nabla u) = 0 \; on \; \Omega \subset \mathbb{R}^2, \\ \sigma\frac{\partial u}{\partial n} = J \; on \; \partial\Omega, \\ \int_{\partial\Omega} u \, ds = 0. \end{cases} \tag{1}$$

Where
u : The electric potential distribution
σ: Given conductivity distribution inside the domain
J: Current injected through the boundary
n: The external normal

3.1.1 Finite Element Method (FEM)

The Finite Element Method is a common choice for solving the forward problem In order to build a finite element model, the division of the domain of study into elements is essential.

An approximation function is used to represent the distribution of the potential u within the element. The concept is to multiply the equation by a test function v and integrate by parts Considering the weak formulation of this problem [10]

$$u \in H_0^1(\Omega) / \forall v \in H_0^1(\Omega) \int_{\Omega} \sigma\nabla u.\nabla v d\Omega = \int_{\Gamma} J v \, d\Gamma \tag{2}$$

A solution to the forward problem can be found by solving the linear system by combining all elements, the resulting formulation is given by

$$[Y(\sigma)] V = I, \tag{3}$$

Where $[Y(\sigma)]$ is the global conductivity matrix, σ is the element conductivity vector, V is the nodal electric potential vector and I is the nodal current vector [11].

3.1.2 Numerical Simulations of Forward Problem

Using the MATLAB PDE toolbox, we can create the geometry of the thorax as well as its simple and refined mesh.

The resulting distribution of the electric potential is shown in Fig. 2:

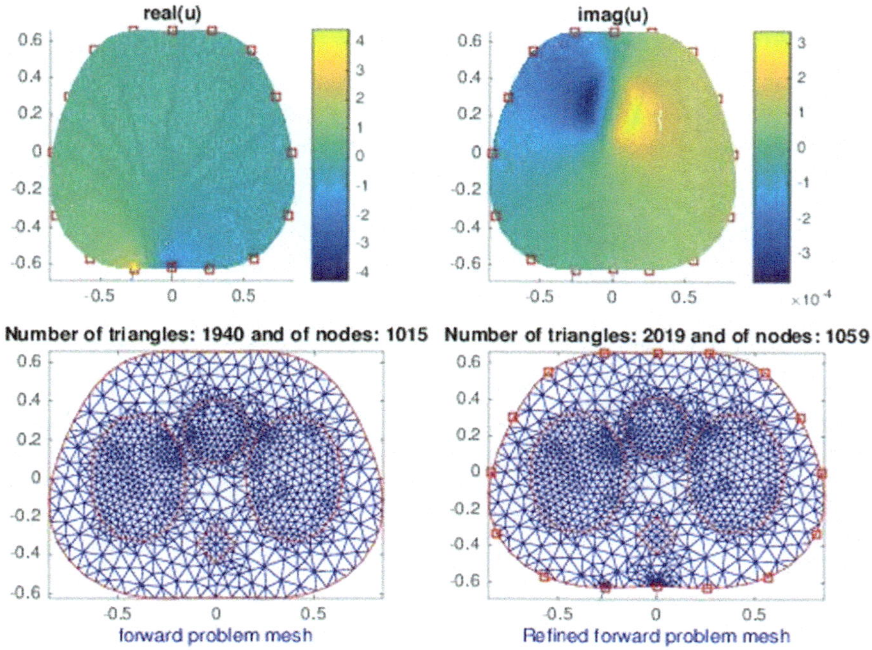

Fig. 2 The refined mesh and the complex value of the distribution of potential

4 The Inverse Problem

4.1 The Mathematical Model of Inverse Problem

An inverse problem is a situation in which the values of certain (unknown) parameters of a model must be identified from observations (measurements) of the phenomenon. It is also the opposite of a direct problem, otherwise an inverse problem consists of determining causes with effects [12].

The EIT is an inverse and ill-posed problem because there is not only a single solution for the reconstruction of images for the given boundary potential distribution (Fig. 3).

Therefore, different methods are adapted to solve the EIT as an optimization problem, the standard method is the use of the iterative regularized Gaussien-Newton method [13]

$$\sigma_{k+1} = \sigma_k + \left(\left(f^{'}(\sigma_k)\right)^T * f^{'}(\sigma_k) + \theta I\right)^{-1} \left(f^{'}(\sigma_k)\right)^T (f(\sigma_k) - V_m) \quad k = 0, 1, 2 \ldots \tag{4}$$

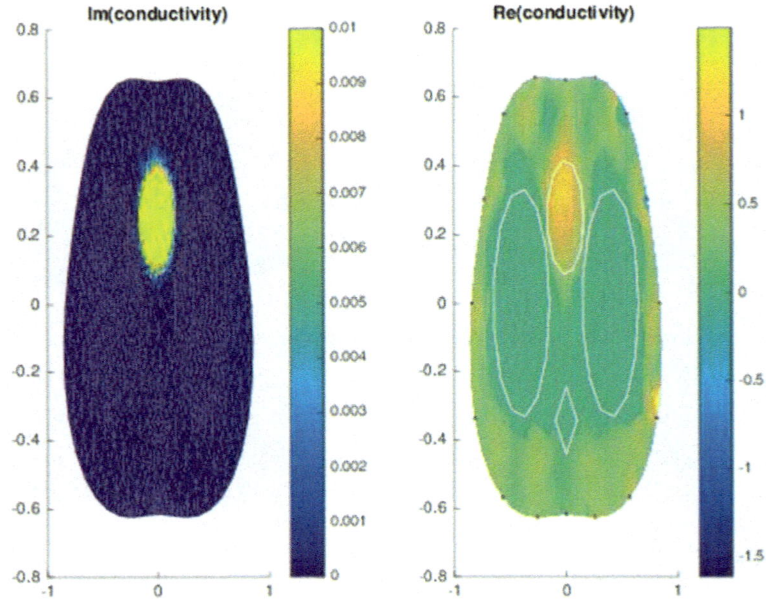

Fig. 3 A two-dimensional model of the complex-valued of given conductivity of the human thorax including lungs, heart and bone

Where $f(\sigma_k)$ is the sensitivity matrix of dimension m x r , m the number of measured voltages I is the identity matrix of size r x r, V_m the measured voltage and θ is a regularization parameter initially fixed at 0.01.

4.2 The Numerical Simulations of Inverse Problem

In this section, we perform numerical simulations to demonstrate our theoretical results . We wrote a code in MATLAB based on an iterative algorithm.
At the end of an iteration, if the value of the objective function of potential has decreased compared to the previous iteration. θ is decreased by a factor of 10, while if the value of the objective function has increased. θ is increased by a factor of 10.

We use data of the Forward problem and the corresponding conductivities of the tissues are taken out of medical literature [13].

The conductivity distribution of the model thorax tissue can be identified from the color in the reconstructed images as shown in the Fig. 4.

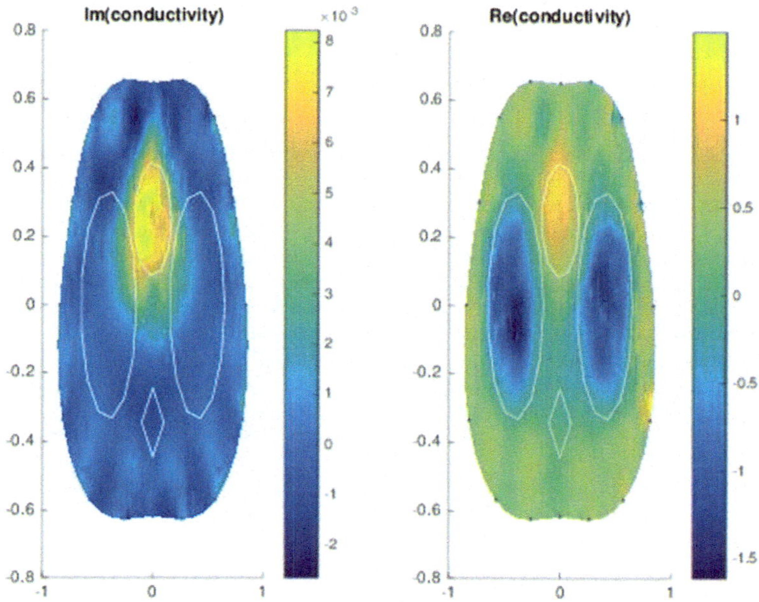

Fig. 4 A two-dimensional model of the complex-valued of reconstructed conductivity of the human thorax including lungs, heart and bone

5 Conclusion

There are several clinical applications that could benefit from EIT technology as a technique for measuring lung function In this paper, the question was approached from two angles: from a theoretical point of view, image reconstruction algorithms were developed and from a numerical point of view, simulations under MATLAB. The most important improvement in the measurement of lung function are:

- Complex impedance distribution
- Image reconstruction algorithms on 3D models.

References

1. Lyazidi, A., Richard, J.C., Dellamonica, J., et al.: Nouvelles perspectives dans le monitoring respiratoire. Réanimation 21, 9–19 (2012)
2. Webster, J.G.: Electrical Impedance Tomography. Adam Hilger Series of Biomedical Engineering, Adam Hilger, New York, USA (1990)
3. Hadinia, M., Jafari, R.: An element-free Galerkin forward solver for the complete-electrode model in electrical impedance tomography. Elsevier (2015)

4. Adler, A., Lionheart, W.R.B.: Uses and abuses of EIDORS: an extensible software base for EIT. IOPscience, Ottawa, Canada (2006)
5. Hikmah, I., Rubiyanto, A., Endarko.: Two-dimensional electrical impedance tomography (EIT) for characterization of body tissue using a gauss-newton algorithm. IOP Conf. Ser. J. Phys. Conf. Ser. **1248** (2019)
6. Kilic, B.: Impedance image reconstruction with artificial neural network in electrical impedance tomography. Eur. J. Tech. (EJT) (2019)
7. Thiago de Castro Martins aAndré Kubagawa Sato aFernando Silva de Moura bErick Dario León Bueno de Camargo bOlavo Luppi Silva bTalles BatistaRattis Santos cZhanqi Zhao d eKnut Mõeller dMarcelo Brito Passos Amato fJennifer L. Mueller gRaul Gonzalez Lima cMarcos de Sales Guerra Tsuzuki a (2019). A review of electrical impedance tomography in lung applications: Theory and algorithms for absolute images, Elsevier
8. Demidenko, E.: An analytic solution to the homogeneous EIT problem on the 2D disk and its application to estimation of electrode contact impedances. IOP Sci. J. (2011)
9. Crabb, M.G.: Convergence Study of 2D Forward Problem of Electrical Impedance Tomography with High-order Finite Elements. Taylor & Francis Online (2016)
10. Lionheart, W.R.B.: EIT reconstruction algorithms: pitfalls, challenges and recent developments. IOP Sci. J. (2004)
11. Silva, O.L., Lima, R.G.: Numerical convergence of 3D electrode models used in electrical impedance tomography. ScienceDirect (2018)
12. Alsaker, M., Mueller, J.L., Murthy, R.: Dynamic optimized priors for d-bar reconstructions of human ventilation using electrical impedance tomography. J. Comput. Appl. Math. (2018)
13. Helfrich-Schkarbanenko, A., Kreutzmann, T., Schmitt, S., Hettlich, F.: A data transformation method in electrical impedance tomography. In: Conference on Applied Inverse Problems Vienna, Austria (2009)
14. Gagnon, H.: Application de la tomographie d'impédance électrique á la résolution du problème inverse en électrocardiographie. École Polytechnique de Montréal (1997)

On Local and Global Bisection-Type Mesh Refinements in C Programming Language

Zhor Mellah and El Bekkaye Mermri

Abstract The finite element method (FEM) is a numerical method of resolution of many problems modeled in terms of partial differential equations. It is a powerful and widely used method in science and engineering. Its simulation requires in a first phase the construction of a mesh of the computational domain. In our paper, we present an efficient approach for local and global mesh refinements of triangular and quadrilateral two-dimensional meshes in C programming language. The proposed refinement algorithms are based on a bisection-type method which produces nested refinements of the triangulation. The algorithms are short, easy to understand and modify, moreover they can be easily integrated in a FEM C program.

1 Introduction

Meshes are used in many application areas, and they are essential for the computation of the numerical solutions of partial differential equations (PDEs). Obviously, the numerical simulation by finite element method (FEM) of many mathematical, physical, and financial problems formulated in terms of PDEs requires, at the first step, the construction of a mesh of the computational domain. This mesh is based on a geometric subdivision of the problem domain into a number of small polygonal subdomains, called elements. Usually, these subdomains are chosen of simple shapes, such as triangles or quadrilaterals in two dimensions and tetrahedra or hexahedra in three dimensions. In this paper, we deal only with meshes composed of triangular and/or quadrilateral elements.

In addition, if a particular approximate solution is deemed too inaccurate, improvement may be obtained by selecting a larger subspace, by using either mesh refinement,

Z. Mellah (✉) · E. B. Mermri
Department of Mathematics, Faculty of Sciences, University Mohammed Premier, 60050 Oujda, Morocco
e-mail: z.mellah@ump.ac.ma

E. B. Mermri
e-mail: e.mermri@ump.ac.ma

S. Melliani et al. (eds.), *Applied Mathematics and Modelling in Finance, Marketing and Economics*, Studies in Computational Intelligence 1114,
https://doi.org/10.1007/978-3-031-42847-0_19

called *h-refinement*, or higher-order basis functions called *p-refinement*. In the case of *h-refinement*, there is a tendency to refine the mesh near the regions of interest, for example in the regions where one thinks that the variation of the solution is large; however, care must be taken to have elements closer to a regular polygon. The more the mesh is narrowed, the more the computed solution will be accurate and close to the exact solution of the problem.

Bisection-type algorithms are very convenient for refining triangular meshes, which is needed for many practical problems, see for instance the works of Rivara [7–9]. This technique became more popular within the FEM community for mesh refinement/adaptation purposes, and several local and global refinement algorithms based on bisection-type algorithms were established, see for instance [3–6]. The refinement is said to be *local* if the partition is carried out on a subset of elements, producing so-called *adaptive refined meshes*. The *global*, also known as *uniform refinement*, concerns the partition of all the elements in a mesh. Bisection-type algorithms guarantee fine-quality structuring with irregular and nested triangulations. To ensure the shape regularity of the elements is inherited by elements created during the refinement process, we restrict attention to bisection-type mesh refinement. A triangular element is subdivided into four geometrically similar triangles to the parent triangle.

In this paper, we present how to proceed to obtain efficient algorithms for the C program for local and global mesh refinement for two-dimensional mesh, as well as simple mesh generation algorithms for rectangular domains. The procedure is based on a bisection-type mesh refinement. We consider triangular and/or rectangular meshes of a polygon. The algorithms are short, easy to understand and modify, moreover they can be integrated in a FEM C program.

The paper is structured as follows. First, we give a brief section about mesh background. Then, in Sect. 3, we present mesh generation algorithms for rectangular domains. Section 4 is devoted to the procedure for refining different meshes of polygonal domains; we consider local and global mesh refinements for triangular and/or rectangular elements. Finally, in Sect. 5, we present the algorithms corresponding to the mesh refinement procedures present in the previous section.

2 Background

A mesh is a geometric data structure that can be used to represent a surface subdivision using a set of polygons and a three-dimensional domain subdivision by a set of volume shapes. It consists of a set of vertices, connected to each other by edges or polygons. The objective of a mesh is to simplify a system by a model representing this system and, possibly, its environment, for the purpose of scientific computations or graphical representations. A mesh can be characterized by its dimension (one, two, or three dimensions), its fineness, the geometry of the elements, and the degree of the element (degree of the polynomial interpolating the vertices of each element).

In the FEM, a mesh is defined by its coordinate system which includes elements and vertices. The vertices are represented by points of the plane or space. In two dimensions, the elements are usually triangles or quadrilaterals. In three dimensions, it is also possible to use voluminal shapes to represent the elements, such as tetrahedrons, hexahedrons, and prisms. In this paper, we only consider two-dimensional domains. A data structure representing a mesh must store several types of elements: finite elements, vertices, and boundary edges. There are several possibilities to represent the meshes, each one having its advantages and its disadvantages. The choice is made in terms of memory occupancy, topological query (browse the neighbors of a vertex, etc.), and ease of modification (insertion/deletion of elements). A two-dimensional mesh generator produces a triangulation starting from an input of a planar straight-line graph (PSLG). A PSLG is a set of vertices and segments. The mesh generation process necessarily divides each segment into smaller edges called sub-segments. In general, a mesh should satisfy some constraints: the union of the triangles is the triangulation domain and the triangles should be relatively "round" in shape, a lower bound on the smallest angle of a triangulation implicitly bounds the largest angle. A *structured* mesh is one in which all interior vertices are topologically alike. An *unstructured* mesh is one in which vertices may have arbitrarily varying local neighborhoods.

Delaunay refinement is a technique for generating unstructured meshes of triangles. The problem is to find a triangulation that covers a specified domain, by maintaining a constrained Delaunay triangulation. The triangulation is refined by inserting carefully placed vertices until the mesh meets the constraints on the triangle quality and size. The angles should not be too small or too large, and the triangles should not be much smaller than necessary, nor larger than desired. The Delaunay triangulation of a point set has the desirable property that it maximizes the minimum angle. For constrained Delaunay triangulations, see [2, 12].

Refinement algorithms based on the longest-edge bisection of triangles were developed to deal with adaptive discretizations. These algorithms guarantee fine-quality structuring with irregular, nested triangulations, mainly as a result of the "bounded" characteristics of the small angles of the triangles thereby created. Such refinement algorithms when applied iteratively to an initial mesh produce a sequence of nested meshes suitable for multi-grid techniques and hierarchical data structures [7, 10].

Mesh refinement techniques are commonly used to solve PDEs with the finite volume method (FVM) or the FEM. The FVM is typically used to solve fluid-flow problems by defining a control volume surrounding each vertex in a mesh and measuring the flux entering and exiting the control volume. Delaunay meshes are extensively used in the FVM because it is easy to define a control volume around a vertex using a Delaunay mesh and its corresponding Voronoï diagram [11]. On the other hand, the FEM may be used with any mesh and they are no more complicated on unstructured meshes than on structured meshes. Furthermore, there is no real advantage the angles meet the bound constraint. Babuška and Aziz [1] showed that in the approximation by FEM the minimum angle condition is not essential, assuming that angles are bounded away from π, a strictly weaker condition.

Adaptive mesh refinement places more grid points in areas where the error in the solution is known or suspected to be large. Local error estimates based on a solution computed on an initial mesh are known as a posteriori error estimates and can be used to determine which elements should be refined. One approach to mesh refinement iteratively inserts extra vertices into the triangulation, typically at edge bisectors or triangle circumcenter.

3 Mesh Generation

In this section, we present some algorithms that can be coded in C language to produce uniform two-dimensional triangular and/or rectangular meshes of simple domains such as a rectangle or a polygon, that can be in the FEM context.

3.1 Mesh Data Structure

Each of the mesh codes provides three lists of data containing the all necessary information for the description of the mesh, recorded in three different files: a file which contains the nodes of the mesh, the second contains the triangular or rectangular elements of the mesh, and the third one contains the segments in the boundary of the domain.

- *File nodes*: Each line contains the two coordinates of a mesh node.
- *File elements*: Each line contains the three (or four) labels of vertices of each triangular (or rectangular) element of the mesh, placed in an anti-clockwise order form left to right.
- *File segments*: Each line contains two labels of the nodes in the extremities of a segment in the boundary of the domain.

Files formats are as follows:

node.txt		element3.txt			element4.txt				segment.txt	
x_1^1	y_1^1	n_1^1	n_2^1	n_3^1	n_1^1	n_2^1	n_3^1	n_4^1	n_1^1	n_2^1
x_1^2	y_1^2	n_1^2	n_2^2	n_3^2	n_1^2	n_2^2	n_3^2	n_4^2	n_1^2	n_2^2
$\vdots$	$\vdots$	$\vdots$	$\vdots$	$\vdots$	$\vdots$	$\vdots$	$\vdots$	$\vdots$	$\vdots$	$\vdots$
x_1^{nn}	y_1^{nn}	n_1^{nte}	n_2^{nte}	n_3^{nte}	n_1^{nre}	n_2^{nre}	n_3^{nre}	n_4^{nre}	n_1^{ns}	n_2^{ns}

where "nn" is the number of the mesh nodes, "nte" and "nre" are the numbers of the triangular and rectangular elements, respectively, and "ns" is the number of the segments.

3.2 Rectangular and Triangular Mesh Generation of a Rectangle

We consider a rectangle R having vertices: (x_1, y_1), (x_2, y_2), (x_3, y_3), and (x_4, y_4), where (x_1, y_1) is its bottom-left corner. For the seek of simplicity, we assume that the sides of the rectangle are parallel to (Ox) and (Oy) axes and we denote by a and b the length of the parallel side to (Ox) and (Oy) respectively. We subdivide a into n subdivisions and b into m subdivisions. The number of nodes is then $(n + 1)(m + 1)$ node, the number of segments is $2(n + m)$ segment and the number of elements is nm element for rectangular mesh and $2nm$ for triangular mesh. The instructions followed in our programs to build the mesh respect the following rules:

- The nodes of the mesh are constructed and numbered from up to down and from left to right starting by the vertex (x_4, y_4).
- The rectangular elements are constructed from up to down and from left to right starting by the left side of the rectangle. The vertices of each element are set in anti-clockwise direction.
- The triangular elements are constructed from up to down and from left to right starting from the left side of the rectangle by splitting each rectangular element from its diagonal into two triangle elements. The vertices of each element are set in anti-clockwise direction, see Fig. 1.
- The segments are built starting by those on the right side of the rectangle, then those on the bottom side, then the left side, and finishing with the top side.

After executing the code, we obtain the following files:

- A *nodes.txt* file containing the coordinates of each node of the mesh.

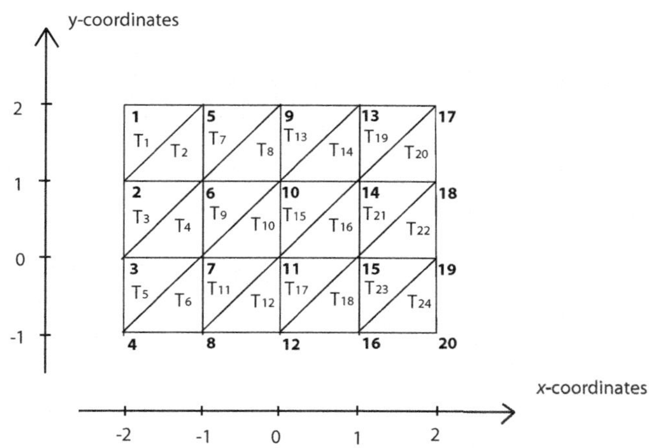

Fig. 1 Example of triangular (rectangular $R_j = T_i \cup T_{i+1}$) mesh

- A *elements3.txt* file that contains the labels (numbers) of the three nodes that define each triangular element of the mesh.
- A *elements4.txt* file that contains the labels of the four nodes that define each rectangular element of the mesh.
- A *segments.txt* file containing the labels of the nodes that define each segment of the mesh at the border of the domain.

The triangular and rectangular meshing algorithm of a rectangle R follows the following steps:
Step 1. Read (x_1, y_1) and (x_3, y_3).
Step 2. Compute (a, b) the length and the width of the rectangle.
Step 3. Compute and add mesh nodes in a file.
Step 4. Compute and add elements into a file.
Step 5. Compute and add mesh segments in a file.

The fragments of C codes to create the mesh are given as follows. First, we define the following structures and variables:

```
typedef struct node{
     float x,y;
     }Node;
typedef struct element3{
     int n1,n2,n3;
     }Element3;
typedef struct element4{
     int n1,n2,n3,n4;
     }Element4;
typedef struct Segment{
     int n1,n2;
     }Segment;

Node *nodes;
Element3 *elements3;
Element4 *elements4;
Segment *segments;
```

Construction of nodes:

```
for(i=0;i<=n;i++){
     for(j=0;j<=m;j++){
                 if(j==0||j==m||i==0||i==n){t=1;}
                          else{t=0;}
                          nodes[j+(m+1)*i].x = x_1+i*(a/n);
                          nodes[j+(m+1)*i].y = (y_1+b)-j*(b/m);
                          nodes[j+(m+1)*i].s = t;
                 }
     }
```

Construction of rectangular elements:

```
k=0;
for(i=0;i<n;i++){
    for(j=0;j<m;j++){
        elements4[k].n1 = j+(m+1)*i+1;
        elements4[k].n2 = j+(m+1)*i+2;
        elements4[k].n3 = j+(m+1)*(i+1)+2;
        elements4[k].n4 = j+(m+1)*(i+1)+1;
        k++;
        }
    }
```

Construction triangular elements:

```
k=0;
for(i=0;i<n;i++){
    for(j=0;j<m;j++){
                        elements3[k].n1 = j+(m+1)*i+1;
                        elements3[k].n2 = j+(m+1)*i+2;
                        elements3[k].n3 = j+(m+1)*(i+1)+1;
                        k++;
                        elements[k]3.n1 = j+(m+1)*i+2;
                        elements[k]3.n2 = j+(m+1)*(i+1)+2;
                        elements[k]3.n3 = j+(m+1)*(i+1)+1;
                        k++;
                        }
    }
```

Construction of segments:

```
for (i=0; i<m; i++) {
     segments[i].n1 = i;
     segments[i].n2 = i+1;
     }
for (i=m; i<m+n; i++) {
     segments[i].n1 = m+(i-m)*(m+1);
     segments[i].n2 = m+(i+1-m)*(m+1);
     }
for (i=m+n; i<2*m+n; i++) {
     segments[i].n1 = (m+1)*(n+1)-1-i+m+n;
     segments[i].n2 = (m+1)*(n+1)-1-(i+1)+(m+n);
     }
for (i=2*m+n; i<2*m+2*n; i++) {
     segments[i].n1 = (m+1)*(n)-(i-(2*m+n))*(m+1);
     segments[i].n2 = (m+1)*(n)-(i+1-(2*m+n))*(m+1);
     }
```

4 Mesh Refinement

We consider the h-refinement of a given mesh. To carry out this refinement, one needs an input of the files containing the description of the initial mesh: elements, nodes coordinates, and segments describing the boundary.

There are two types of refinement: global and local refinements. We perform the refinement on rectangular and triangular or mixed meshes.

4.1 Global Refinement of a Rectangular, Triangular, and Mixed Meshes

To perform the global refinement of any mesh, we cross the list of all elements of the initial mesh. We refine each element by adding new local nodes to this element, which allows us to create new elements (called children) and new segments. The new nodes are added to the node file, and the new elements and segments come in the palace of the elements and segments of the initial mesh. The addition of the new local nodes is as described as follows:

- For quadrilateral elements of vertices n_i, $i = 1, \ldots, 4$, we put nodes on the middle points of each side of the element and one in the center of the quadrilateral rectangle. The new elements are quadrilaterals obtained by connecting the middle point N_5 with the new nodes (N_i, $i = 1, \ldots, 4$), see Fig. 2. Nodes of each new element are set in the file in anticlockwise order, which is desirable in the finite element computation. The four new elements will replace the initial one.

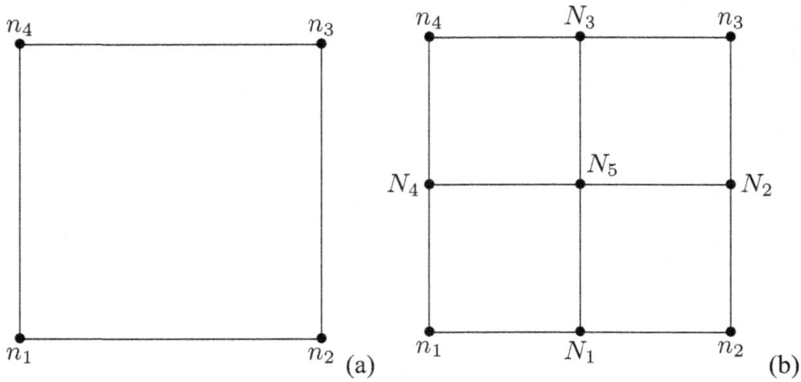

Fig. 2 Refinement of a rectangular element: **a** element before refinement; **b** element after refinement

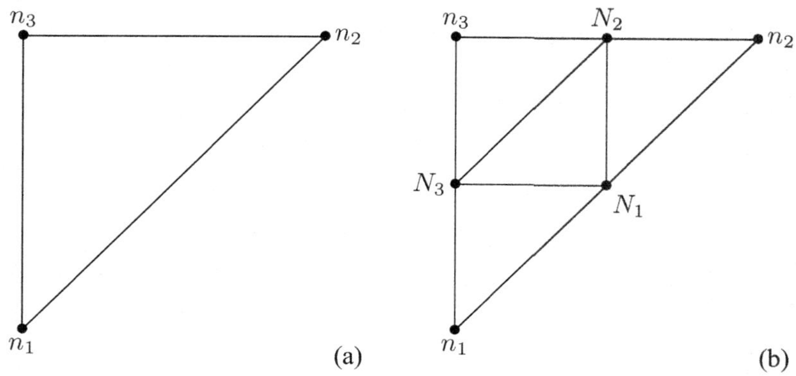

Fig. 3 Refinement of a triangular element: **a** element before refinement; **b** element after refinement

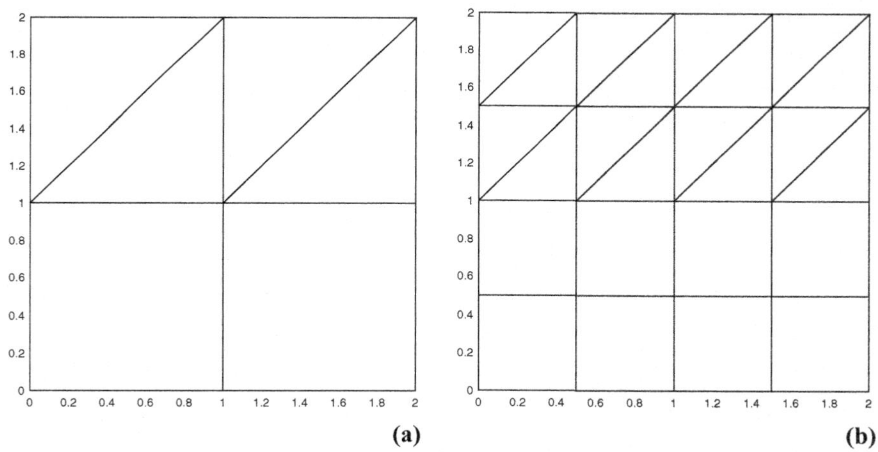

Fig. 4 A global refinement for a mixed mesh: **a** a mixed mesh before refinement; **b** the mixed mesh after global refinement

- For triangular elements of vertices n_i, $i = 1, 2, 3$, we put nodes on the middle points of each side of the triangle. The four new elements are triangles obtained by connecting the new nodes (N_i, $i = 1, 2, 3$) with each other, see Fig. 3. We remark that the new triangles are similar to the original one, that is they have the same angles as the parent triangle. Nodes of each new element are set in anticlockwise order. The new elements will replace the initial one.

Figure 4 illustrates an example of global refinement of a mixed mesh (with triangular and rectangular elements). Algorithms for global refinement of a triangular mesh are given by Algorithms 1 and 2 given in Sect. 5.

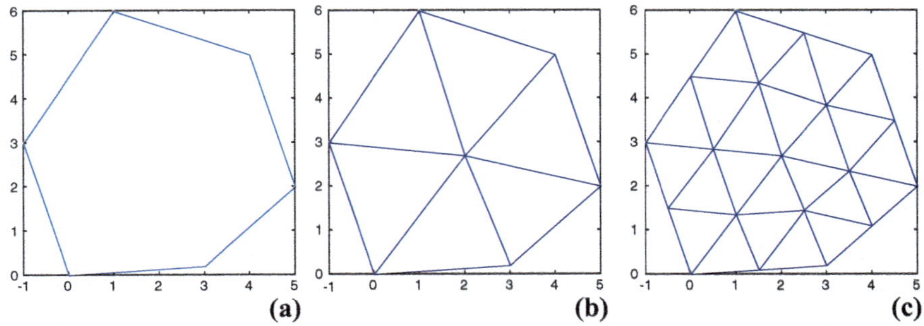

Fig. 5 Meshing a polygonal domain by refinement: **a** polygon to mesh; **b** initial triangular mesh of the polygon; **c** the polygon mesh after refinement

Remark 1 We remark that one can get an initial mesh of a polygonal domain by computing its center, then connecting it to the external nodes of the polygon as is shown in Fig. 5. Then, by refining this initial mesh, we obtain a new mesh of the polygon. We note that the smallest rectangle containing the polygon must not be very narrow, in order to avoid small angles in the elements.

4.2 *Local Refinement of a Triangular Mesh*

The local refinement of a mesh is carried out by refining some given elements of this mesh. To perform this refinement, it is necessary to have a file containing the elements or construct it by using the set of nodes defining the region to be refined. First, we take out the elements to be refined for the list of elements of the mesh. Next, we refine the elements in question in the same way as already described in the previous section. The new mesh is composed of the list of elements of the mesh private from elements to be refined and the list of refined elements. In the end we may have a non-conforming mesh where some elements my have additional nodes in their edges, see left side figures in Figs. 6, 7 and 8. To solve this problem, we re-mesh the neighboring elements which cause this non-conforming mesh problem; there are three scenarios:

1. If the element contains only one node which causes the problem, we connect it to the opposite vertex to split the element into two elements which will replace the parent element in the mesh, see Fig. 6.
2. If the element contains two nodes that cause the problem, we connect them and connect one of them to the opposite vertex in order to decompose the element in question to three elements. Then the new tree elements will replace the element causing the problem, see Fig. 7.

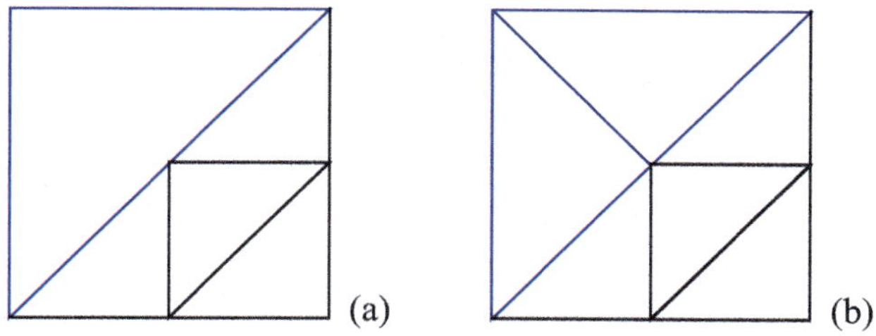

Fig. 6 First scenario to refine non-conforming triangular elements: **a** element before refining non-conforming element; **b** element after refining non-conforming element

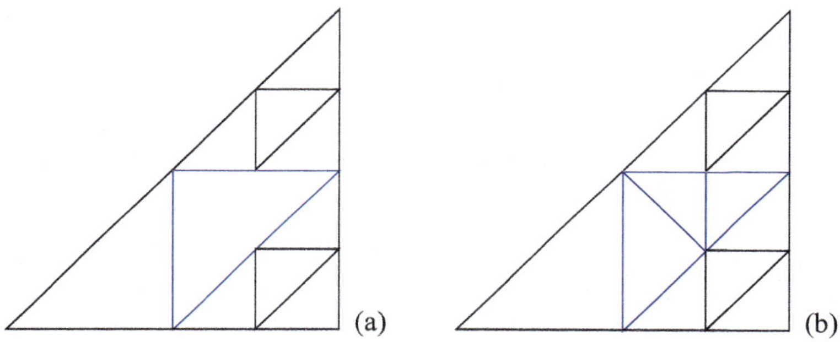

Fig. 7 Second scenario to refine non-conforming triangular elements: **a** element before refining non-conforming element; **b** element after refining non-conforming element

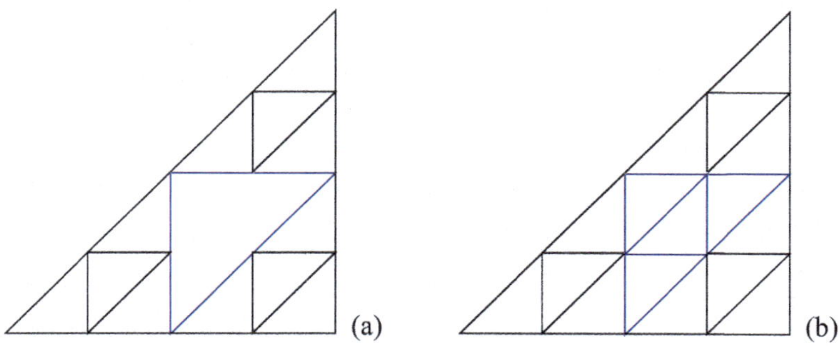

Fig. 8 Third scenario to refine non-conforming triangular elements: **a** element before refining non-conforming element; **b** element after refining non-conforming element

3. If the element contains three nodes that cause the problem, we connect them in the same way as we carried out the refinement of an element, see Fig. 3. Then we get four new elements which will replace the element causing the problem, see Fig. 8.

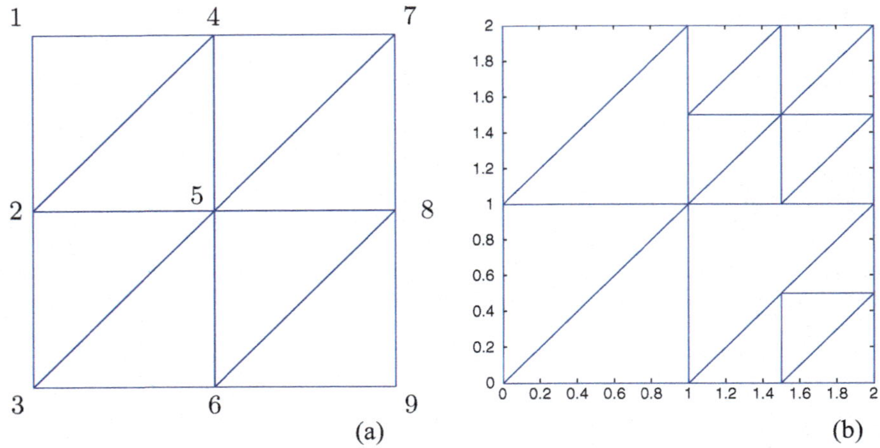

Fig. 9 Triangular mesh before (**a**) and after (**b**) local refinement of the elements containing nodes number 7 and 9

Fig. 10 Conforming triangular refined mesh

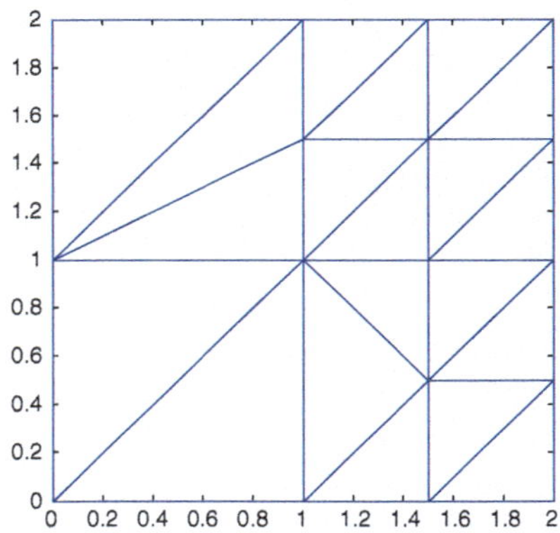

As an example, we consider the mesh in the left-hand side of Fig. 9, where we want to refine the elements around nodes labeled 7 and 9. When we refine locally this mesh, we get the non-conforming mesh on the right-hand side of the same figure. The corrected refined mesh is given by Fig. 10.

5 Refinement Algorithms

Data of the mesh: nodes, elements, and segments are represented by the following structures:

```
structure node{
      x, y: real;
      label: integer;
}
structure element3{
      n1, n2, n3: integer;
}
structure element4{
      n1, n2, n3, n4: integer;
}
structure segment{
     n1, n2: integer;
}
```

We define the following variables:

$A[N, N]$	: array of **reals**;
$b[N]$	: vector of **reals**;
$v[3]$	: vector of **integers**; (or $v[4]$ for rectangular elements)
element[L]	: array of **structure element3**; (or element4)
Relement[ ]	: array of **structure element3**; (or element4)
nodes[N]	: array of **structure node;**
segment[Ns]	: array of **structure segment**;
Rsegment[ ]	: array of **structure segment**;

where **N** and **L** are, respectively, the number of nodes and elements in the mesh; **Ns** is the number of segments in the boundary.

- The elements and segments of the initial mesh are, respectively, stored in the arrays *element*[ ] and *segment*[ ].
- The elements and segments of the refined mesh are, respectively, stored in the arrays *Relement*[ ] and *Rsegment*[ ].
- The matrix A is used to store the labels (numbers) of new nodes in the refined mesh; if $A[i][j] = 0$ then the segment $[i, j]$ has no new node, otherwise the segment contains a new node of label $N = A[i][j]$.

5.1 Algorithms for Global Refinement

Algorithm 1 Algorithm for global refinement of a triangular mesh

```
for i ← 1 to N do
  for j ← 1 to N do
    A[i][j] ← 0
  end for
end for
k ← 0
for i ← 1 to L do
  i1 ← element[i].n1
  i2 ← element[i].n2
  i3 ← element[i].n3

  N1.x ← (node[i1].x + node[i2].x)/2          ▷ Creates the nodes N1, N2, N3
  N1.y ← (node[i1].y + node[i2].y)/2
  N2.x ← (node[i2].x + node[i3].x)/2
  N2.y ← (node[i2].y + node[i3].y)/2
  N3.x ← (node[i3].x + node[i1].x)/2
  N3.y ← (node[i3].y + node[i1].y)/2

                                              ▷ Adds the new nodes to the list of nodes
  if A[i1][i2] = 0 then                       ▷ Means that N1 is not yet
                                                added to the list of nodes
    N ← N + 1
    node[N] ← {N1.x, N1.y}
    A[i1][i2] ← N
    A[i2][i1] ← N
  end if
  if A[i2][i3] = 0 then
    N ← N + 1
    node[N] ← {N2.x, N2.y}
    A[i2][i3] ← N
    A[i3][i2] ← N
  end if
  if A[i3][i1] = 0 then
    N ← N + 1
    node[N] ← {N3.x, N3.y}
    A[i3][i1] ← N
    A[i1][i3] ← N
  end if
                                              ▷ Creates the list of refined elements
  j1 ← A[i1][i2]
  j2 ← A[i2][i3]
  j3 ← A[i3][i1]

  Relement[k + 1].n1 ← {i1, j1, j3}           ▷ Adds (n1,N1,N3)
  Relement[k + 2].n1 ← {j1, i2, j2}           ▷ Adds (N1,n2,N2)
  Relement[k + 3].n1 ← {j2, i3, j3}           ▷ Adds (N2,n3,N3)
  Relement[k + 4].n1 ← {j1, j2, j3}           ▷ Adds (N1,N2,N3)

end for
```

Algorithm 2 Algorithm for refining segments

```
1: k ← Ns
2: for i ← 1 to Ns do
3:    i1 ← S[i].n1
4:    i2 ← S[i].n2
5:    j ← A[i1][i2]
6:    k ← k + 1
7:    Rsegment[k] ← {i1, j}
8:    Rsegment[k + 1] ← {j, i2}
9:    k ← k + 1
10: end for
```

5.2 *Algorithms for Local Refinement*

In addition to the notations given in the beginning of this section, we use the following arrays:

—**Lelement**[Le] to store the set of elements to be refined.
—**NCelement**[NC] to store the non-conforming elements produced after the refinement process.

6 Conclusion

In this paper, we have presented some algorithms for two-dimensional mesh refinement that can be implemented in C/C++ programming language or any other compiled programming languages. We have considered global and local refinement of a triangular and/or rectangular meshes of polygonal domains. We have also given a simple triangular and rectangular mesh generation programs for rectangular domains.

Our aim was to give programs that are short, easy to understand and modify, and can be integrated in a FEM C program.

Algorithm 3 Algorithm for local refinement of a triangular mesh

```
for i ← 1 to N do
  for j ← 1 to N do
    A[i][j] ← 0
  end for
end for
k ← 0
for i ← 1 to Le do
  i1 ← Lelement[i].n1
  i2 ← Lelement[i].n2
  i3 ← Lelement[i].n3

  N1.x ← (node[i1].x + node[i2].x)/2          ▷ Creates the nodes N1, N2, N3
  N1.y ← (node[i1].y + node[i2].y)/2
  N2.x ← (node[i2].x + node[i3].x)/2
  N2.y ← (node[i2].y + node[i3].y)/2
  N3.x ← (node[i3].x + node[i1].x)/2
  N3.y ← (node[i3].y + node[i1].y)/2

                                              ▷ Update the list of nodes
  if A[i1][i2] = 0 then                       ▷ Means that N1 is not yet added
    N ← N + 1                                   to the list of nodes
    node[N] ← {N1.x, N1.y}
    A[i1][i2] ← N
    A[i2][i1] ← N
  end if
  if A[i2][i3] = 0 then
    N ← N + 1
    node[N] ← {N2.x, N2.y}
    A[i2][i3] ← N
    A[i3][i2] ← N
  end if
  if A[i3][i1] = 0 then
    N ← N + 1
    node[N] ← {N3.x, N3.y}
    A[i3][i1] ← N
    A[i1][i3] ← N
  end if
  j1 ← A[i1][i2]
  j2 ← A[i2][i3]
  j3 ← A[i3][i1]
                                              ▷ Update the list of refined elements
  Relement[k + 1] ← {i1, j1, j3}              ▷ Adds (n1,N1,N3)
  Relement[k + 2] ← {j1, i2, j2}              ▷ Adds (N1,n2,N2)
  Relement[k + 3] ← {j2, i3, j3}              ▷ Adds (N2,n3,N3)
  Relement[k + 4] ← {j1, j2, j3}              ▷ Adds (N1,N2,N3)

  k ← k + 4
end for
```

Algorithm 4 Algorithm for updating "*Relement[ ]*" and re-mesh non-conforming elements

```
                                          ▷ Let k be the number of elements in of Relement[ ]
for i ← 1 to L do
  i1 ← element[i].n1                      ▷ Non-conforming element
  i2 ← element[i].n2
  i3 ← element[i].n3

  a12 ← A[i1][i2]
  a23 ← A[i2][i3]
  a31 ← A[i3][i1]

  if a12 = 0 and a23 = 0 and a31 = 0 then
    Relement[k + 1] ← {i1, i2, i3}        ▷ Adds the non-refined elements to the list: Relement[ ]
    k ← k + 1
  else                                    ▷ Re-mesh the non-conforming elements
    if a12 <> 0 and a23 = 0 and a31 = 0 then
      Relement[k + 1] ← {a12, i3, i1}
      Relement[k + 2] ← {a12, i2, i3}
      k ← k + 2
    end if
    if a12 = 0 and a23 <> 0 and a31 = 0 then
      Relement[k + 1] ← {a23, i3, i1}
      Relement[k + 2] ← {a23, i1, i2}
      k ← k + 2
    end if
    if a12 = 0 and a23 = 0 and a31 <> 0 then
      Relement[k + 1] ← {a23, i1, i2}
      Relement[k + 2] ← {a23, i2, i3}
      k ← k + 2
    end if
    if a12 <> 0 and a23 <> 0 and a31 = 0 then
      Relement[k + 1] ← {a12, i2, a23}
      Relement[k + 2] ← {a12, a23, i3}
      Relement[k + 3] ← {a12, i3, i1}
      k ← k + 3
    end if
    if a12 <> 0 and a23 = 0 and a31 <> 0 then
      Relement[k + 1] ← {i1, a12, a31}
      Relement[k + 2] ← {a12, i3, a31}
      Relement[k + 3] ← {a12, i2, i3}
      k ← k + 3
    end if
    if a12 = 0 and a23 <> 0 and a31 <> 0 then
      Relement[k + 1] ← {a23, i3, a31}
      Relement[k + 2] ← {a23, a31, i1}
      Relement[k + 3] ← {a23, i1, i2}
      k ← k + 3
    end if
    if a12 <> 0 and a23 <> 0 and a31 <> 0 then
      Relement[k + 1] ← {i1, a12, a31}
      Relement[k + 2] ← {a12, i2, a23}
      Relement[k + 3] ← {a23, i3, a31}
      Relement[k + 4] ← {a13, a23, a31}
      k ← k + 4
    end if
  end if
end for
```

Algorithm 5 Algorithm for refining segments

for $i \leftarrow 1$ **to** Ns **do**
  $i1 \leftarrow \mathrm{S}[i].n1$
  $i2 \leftarrow \mathrm{S}[i].n2$
  if $A[i1][i2] <> 0$ **then**
    $j \leftarrow A[i1][i2]$
    $k \leftarrow k + 1$
    $\mathrm{Rsegment}[k] \leftarrow \{i1, j\}$
    $\mathrm{Rsegment}[k + 1] \leftarrow \{j, i2\}$
  end if
end for

References

1. Babuška, I., Aziz, A.K.: On the angle condition in the finite element method. SIAM J. Numer. Anal. **13**(2), 214–226 (1976)
2. Chew, L.P.: Constrained Delaunay triangulations. Algorithmica **4**(1), 97–108 (1989)
3. Dunning, D., Marts, W., Robey, R.W., Bridges, P.: Adaptive mesh refinement in the fast lane. J. Comput. Phys. **406**, 1–15 (2020)
4. Hannukainen, A., Korotov, S., Křížek, M.: On global and local mesh refinements by a generalized conforming bisection algorithm. J. Comput. Appl. Math. **235**(2), 419–436 (2010)
5. Korotov, S., Křížek, M., Kropáč, A.: Strong regularity of a family of face-to-face partitions generated by the longest-edge bisection algorithm. Comput. Math. Math. Phys. **48**(9), 1687–1698 (2008)
6. Plaza, Á., Falcón, S., Suárez, J.P., Abad, P.: A local refinement algorithm for the longest-edge trisection of triangle meshes. Math. Comput. Simul. **82**(12), 2971–2981 (2012)
7. Rivara, M.C.: Algorithms for refining triangular grids suitable for adaptive and multigrid techniques. Int. J. Numer. Meth. Eng. **20**(4), 745–756 (1984)
8. Rivara, M.C.: Selective refinement/derefinement algorithms for sequences of nested triangulations. Int. J. Numer. Meth. Eng. **28**(12), 2889–2906 (1989)
9. Rivara, M.C., Iribarren, G.: The 4-triangles longest-side partition and linear refinement algorithm. Math Comp. **65**(216), 1485–1502 (1996)
10. Rivara, M.C., Levin, C.: A 3-D refinement algorithm suitable for adaptive and multi-grid techniques. Commun. Appl. Numer. Methods **8**(5), 281–290 (1992)
11. Sastry S.P.: A 2D advancing-front Delaunay mesh refinement algorithm (2018). http://arxiv.org/abs/1808.01539v2
12. Shewchuk, J.R.: Reprint of: Delaunay refinement algorithms for triangular mesh generation. Comput. Geom.: Theory Appl. **47**(7), 741–778 (2014)

GPSR Compliance

The European Union's (EU) General Product Safety Regulation (GPSR) is a set of rules that requires consumer products to be safe and our obligations to ensure this.

If you have any concerns about our products, you can contact us on ProductSafety@springernature.com

In case Publisher is established outside the EU, the EU authorized representative is:

Springer Nature Customer Service Center GmbH
Europaplatz 3
69115 Heidelberg, Germany

The manufacturer's authorised representative in the EU is Springer Nature Customer Service Centre GmbH, Europaplatz 3, 69115 Heidelberg, Germany. If you have any concerns regarding our products, please contact ProductSafety@springernature.com

Printed and bound by CPI Group (UK) Ltd, Croydon, CR0 4YY
28/09/2026
02224085-0002